Rubidium Atomic Clock
The Workhorse of Satellite Navigation

Rubidium Atomic Clock
The Workhorse of Satellite Navigation

G M Saxena
National Physical Laboratory, India

Bikash Ghosal
Space Applications Centre, India

World Scientific

EW JERSEY • LONDON • SINGAPORE • BEIJING • SHANGHAI • HONG KONG • TAIPEI • CHENNAI • TOKYO

Published by

World Scientific Publishing Co. Pte. Ltd.
5 Toh Tuck Link, Singapore 596224
USA office: 27 Warren Street, Suite 401-402, Hackensack, NJ 07601
UK office: 57 Shelton Street, Covent Garden, London WC2H 9HE

British Library Cataloguing-in-Publication Data
A catalogue record for this book is available from the British Library.

RUBIDIUM ATOMIC CLOCK
The Workhorse of Satellite Navigation

Copyright © 2020 by World Scientific Publishing Co. Pte. Ltd.

All rights reserved. This book, or parts thereof, may not be reproduced in any form or by any means, electronic or mechanical, including photocopying, recording or any information storage and retrieval system now known or to be invented, without written permission from the publisher.

For photocopying of material in this volume, please pay a copying fee through the Copyright Clearance Center, Inc., 222 Rosewood Drive, Danvers, MA 01923, USA. In this case permission to photocopy is not required from the publisher.

ISBN 978-981-3279-48-3

For any available supplementary material, please visit
https://www.worldscientific.com/worldscibooks/10.1142/11249#t=suppl

Desk Editor: Cheryl Heng

Typeset by Stallion Press
Email: enquiries@stallionpress.com

*I (GMS) am pleased to dedicate this book to my grand-daughter,
Ishika
(Daughter of Shilpi and Bhavya)
an inspiration for conceptualizing and writing this book*

Preface

Time is the most important part of human lives and their activities. Since time immemorial, Time measurement remained in the central stage of all kinds of measurements and underwent numerous and far reaching changes. At present, different kinds of atomic clocks are being used in telecommunication, digital financial transactions, space explorations and many other applications. The satellite navigation has become integral part of everyone's life, as it helps in navigating to one's destination, even in rainy days or dark nights, in thick forests or advanced metropolis, with a precision of a few centimetres to metres. All this could be possible because of the on-board Rubidium (Rb) atomic clocks. To highlight the accuracy and stability of these Rb atomic clocks, we should know that an inaccuracy of $1/100^{th}$ of a Second may lead to a positional error of 30000 km. So, for an accuracy of a few cms/metres in our day to day GPS guided travel by cars or app based Taxis or in others scientific and commercial activities, the role of these Rb atomic clocks and their accuracy can be appreciated. It is natural to ask how the highly accurate Time ticks of the on-board atomic clocks help in achieving the precision in distance. Simple, at the heart of (Global Navigation Satellite System) GNSS workings, is the fundamental relation among distance, light velocity and time i.e., Distance = Velocity × Time. The Rb atomic clocks have gained the place of pride in satellite navigation, because of their light weight and small size, leaving behind several state-of-the-art atomic clocks. These state-of-the-art atomic clocks could not challenge Rb clock's supremacy in the satellite navigation systems like GPS, GLONASS, GALILEO, COMPASS and IRNSS. Most of the on-board atomic clocks in these satellite navigation systems are

Rb atomic clocks. It is rightly the darling and workhorse of the satellite navigation systems. In spite of being one of the earliest and the oldest atomic clocks, researchers are still working on it, to unfold its mysteries and improve its performance. All this sums up the intense curiosity to explore, what is so special about this clock and as a result bring about the conceptualization of this book. In the book, history of time keeping and chronological development of clocks are included. The Rb atomic clock is described including all its details with special attention on its critical part — the Physics package. The Physics package mainly decides the performance of the clock. Now sit back and enjoy reading about the Rb atomic clock. Special efforts are made to describe the Rb atomic clock in a simple way, to benefit the maximum readership. This is perhaps, the only book of its kind, giving all the details of the Rb atomic clock in one place and one needs not shuffle numerous papers and literature to get abridged information on the clock. All in all, the researchers and the industrialists interested in R&D and producing Rb atomic clocks, as a marketable state-of-the-art product, will find the book very useful and practical. In the last chapter, the recipe for developing/manufacturing the Rb atomic clock is given, that should be quite useful even for those interested in developing the Rb atomic clocks for the first time.

G M Saxena
Bikash Ghosal

Acknowledgement

With a great sense of gratitude, I (GMS) wish to express my regards and appreciation for Prof B.S. Mathur, who introduced me to the subject of optical pumping and Rb atomic clock. The free access to him for the discussions on these complex subjects, is highly appreciated. We are thankful to the present and the former Directors, National Physical Laboratory, New Delhi India and SAC-ISRO, for making it possible to do experimental R&D work on Rb atomic clocks. This enriched our experience in the theoretical and experimental aspects of the Rb atomic clock. The experience gained as the leader (GMS) of CSIR-NPL and SAC-ISRO India collaborative project on Rb atomic clock for IRNSS, helped in understanding the stringencies of space qualification of the Rb atomic clock. Dr Pardeep Mohan, Dr S S Verma, Mr A. Banik, Mr Satender Singh, Ms Savita Singh and SAC-ISRO scientists need special mention for their contribution in the work on the Rb atomic clock. We thankfully, acknowledge the institutions, laboratories where we worked and gained wide-ranging experiences that helped immensely in conceptualizing this book. The unclassified information on internet have been used in this book and we sincerely acknowledge all those web sites and their creators.

Contents

Preface vii

Acknowledgement ix

List of Tables xv

List of Figures xvii

Glossary and Acronyms xxv

Introduction 1
    0.1   Satellite navigation systems — global
           and regional 2

Chapter 1.   Overview of Rb Atomic Clocks and the
                  Space Specifications   10
    1.1   Working principle of Rb atomic clock 10
    1.2   Microwave cavity 23
    1.3   Effect of photodiode inside/outside microwave
           cavity and its characteristics 24
    1.4   Rb lamp 27
    1.5   Frequency synthesizer 28
    1.6   Performance specifications of rubidium atomic
           clock for satellite navigation systems 34

Chapter 2.   Theoretical Aspects of Rb Atomic Clock   41
    2.1   Atomic structure of Rb atom 41
    2.2   Optical pumping 45

2.3 Microwave-optical double resonance 47
2.4 The space qualified Rb clocks 48

Chapter 3. Studies on Rb Lamp and Driver Oscillator   49

3.1 Critical role of Rb lamp 49
3.2 Characterization of RF discharge Rb lamp 51
3.3 Experimental test set-up for electrical
     characterization of Rb lamp 60
3.4 Rb lamp driver circuit design aspects 64
3.5 Configuration simulation and development
     of Rb lamp exciter 67
3.6 Spectral profile of Rb lamp and operating
     temperature vs modes 75

Chapter 4. Thermal Analysis of Rb Physics
            Package Components   86

4.1 Introduction 86
4.2 Thermal control systems 87
4.3 Overall power budget of physics package 87
4.4 Thermal simulations and analysis 88
4.5 Heat dissipation problem in Rb bulb 90
4.6 Heat loss mechanism in physics package
     assembly 98
4.7 Experimental setup for thermo vacuum test 100
4.8 Thermo-vacuum test cycle 101
4.9 Results 104

Chapter 5. RF Synthesizer for Rb Atomic Clock   106

5.1 Introduction 106
5.2 Synthesizer design methodology 107
5.3 Precision oven controlled crystal oscillator 109
5.4 Design and simulation of 10 MHz × 9 active
     frequency multiplier 111

5.5	Design and simulation of 360 MHz × 19 SRD multiplier	130
5.6	Design of two stage 6.8 GHz amplifier	147
5.7	Design of medium and high power 90 MHz and 360 MHz amplifier	154
5.8	Causes of frequency offset	157
5.9	Phase noise measurement	161

Chapter 6. Design Simulation and Development of Microwave Cavity 165

6.1	Introduction	165
6.2	Theoretical analysis of cavity Q	168
6.3	3-D cavity simulations in HFFS	171
6.4	Cavity performance and measurement results	177
6.5	Frequency shifts due to cavity pulling effect	182
6.6	Frequency sensitivity of microwave cavity	184

Chapter 7. Design and Simulation of Analog Lock-in Amplifier 188

7.1	Introduction	188
7.2	Physical description and experimental characterization of discriminator signal	191
7.3	Functional aspects of the lock-in amplifier	195
7.4	Design and implementation of the lock-in amplifier	196
7.5	SNR measurement test setup	206

Chapter 8. Magnetic Shield Assembly for The Rb Physics Package 211

8.1	Introduction	211
8.2	The Rb Physics package and its sensitivity factors	212
8.3	Magnetic shield	214
8.4	Base plate	221

Chapter 9. Integrated Testing and Characterization
of Rb Clock Parameters   223

   9.1   Introduction 223
   9.2   Critical clock parameters characterization 225
   9.3   Analysis of Rb resonance frequency dependence
        on various parameters 231
   9.4   The outer space and radiation effects 235

Chapter 10. Recipe of Rb Atomic Clock   238

   10.1 Introduction 238
   10.2 Rb Physics package 239
   10.3 Rb lamp life span determination by calorimeter . . . 251
   10.4 Electronic package 254

Conclusion   269

Bibliography   270

Index   288

List of Tables

4.1	Estimated power budget for Rb atomic clock.	88
4.2	Warm-up period of different sub-assemblies for IP temperature =5°C.	102
4.3	Power dissipation in lamp assembly.	103
4.4	Heater power estimation during the Thermo-vacuum test.	104
5.1	Performance summary and comparisons of frequency multipliers.	128
5.2	Measured phase noise results at 10 MHz and 90 MHz carrier frequency.	129
5.3	Simulation vs measured results of SRD multiplier circuit.	146
5.4	Summary of simulated and measured RF characteristics in two-stage amplifier.	154
5.5	The measured results of RF Synthesizer.	156
6.1	Loop length and the corresponding S_{11} and f_0.	177
7.1	Measured results of servo lock-in amplifier.	208
8.1	Magnetic shield dimensions.	221

List of Figures

1.1	Energy levels of ^{87}Rb and ^{85}Rb.	11
1.2	The g.s hyperfine components with the ^{87}Rb radiation line-a filtered by the ^{85}Rb line-A overlap, ^{87}Rb radiation line-b available for the optical pumping.	12
1.3	Rb physics package-integrated filter cell.	15
1.4	Rb physics package-separate filter cell.	18
1.5	Block diagram of Rb atomic clock (IFT).	26
1.6	Block diagram of Rb clock (SFT).	27
1.7	Sectional schematic of Rubidium atomic clock.	36
2.1	The energy level diagram of the 5th electron shell of ^{87}Rb for the ground and first excited states.	42
2.2	The magnetic sublevels and allowed transitions in the ground state hyperfine of ^{87}Rb.	43
3.1	LC resonator circuit with Rb bulb.	52
3.2	LC resonator circuit (a) Only L & C (b) L &C with Rb bulb (c) L, C and Sill fill material.	52
3.3	Coil impedance vs frequency plot with and without bulb.	53
3.4	Frequency vs LC impedance plot without bulb (purple) with bulb (black) plus sill fill (green).	54
3.5	Bulb assembly with thermal epoxy and heater.	56
3.6	(a) S_{11} = 2dB only with LC resonator (b) S_{11} = 7dB in LC resonator with bulb (c) S_{11} = 1.5dB in resonator with bulb and sill fill epoxy.	57
3.7	Schematic diagram of ADS simulation setup.	58

3.8	Shows the ADS simulation results of (a) reflected power by the resonant circuit (b) transmitted power by the Rb bulb.	59
3.9	Power absorbed and transmitted by the Rb bulb as a function of operating frequency.	60
3.10	Experimental test setup for Rb lamp measurement system.	61
3.11	The reflected power vs frequency in Rb bulb.	63
3.12	Schematic of common-base Clapp oscillator circuit.	65
3.13	Plot of f_T vs. I_C with different value of VCE.	66
3.14	Schematic of bias tee.	67
3.15	Response of bias tee.	68
3.16	Forward S-parameter S_{21} amplitude for the amplifier-resonator cascade. The gain peaks at 100 MHz at about 2.7 dB.	69
3.17	Linear S-parameter (S_{21}) phase for the amplifier-resonator cascade. The phase is zero at 100MHz.	70
3.18	The complete fabricated oscillator PCB with PCB holder, Rb bulb housing and assembled thermal heater.	71
3.19	Dependence of breakdown voltage on the frequency of the exciter field.	72
3.20	Voltage-current characteristics of the electrode-less (EL) Rb discharge lamp.	73
3.21	Output power spectrum of Clapp Oscillator.	74
3.22	Photo of Rb bulb with 10mm diameter.	77
3.23	Experimental test setup for studying the spectrum of Rb lamp.	78
3.24	The spectrum of Rb lamp along with Xenon lines.	79
3.25	The dependence of the spectral intensity variation the temperature (a) D1 line (b) D2 line (c) Xenon inert gas line (d) Spectrum of D1 & D2 line at 90°C.	81
3.26	(a) The intensity of spectral lines D1, D2 and Xenon (b) Mean count ratio of D1 line with respect to D2 and Xenon line.	82

3.27	Three different transition modes of Rb bulb (a) White mode (b) Ring mode (c) Red mode.	83
3.28	The Rb lamp output (Photodiode current) as a function of time.	84
4.1	Assembled view of developed thermal mathematical model.	89
4.2	TMM details of main assembly.	90
4.3	Temperature distribution on the Rb bulb.	91
4.4	Temperature distribution on Rb bulb assembly.	94
4.5	Temperature distributions (10°C) on absorption cell.	95
4.6	Temperature distribution (°C) on absorption cell assembly.	96
4.7	Temperature distribution on main housing.	97
4.8	Temperature distribution on base plate.	97
4.9	Assembly of physics package for thermo vacuum test.	100
4.10	Thermo-vacuum test cycle during the TV test.	101
4.11	Layout of DVM lamp driver PCB.	102
4.12	Temperature profile for various sub-assemblies during IP = 5°C cycle.	104
5.1	Block diagram of Rb frequency synthesizer.	108
5.2	Phase noise plot for Centum OCXO locked with Cesium source.	110
5.3	Ideal harmonic collector current as a function of BJT collector conduction angle.	116
5.4	Simulated collector current waveforms of 10×9 frequency multiplier.	117
5.5	Simulated output spectrum without idler network.	122
5.6	Simulated idler network response.	123
5.7	Simulated output spectrum of multiplier with idler network.	124
5.8	Fabricated 10MHz ×9 frequency multiplier.	125
5.9	Wideband output spectrum of the 10MHz ×9 Multiplier.	126
5.10	Measured output power (Pout) and conversion gain with respect to input power (Pin).	127

5.11	Measured output phase noise spectrum of the BJT 10 × 9 frequency multiplier.	129
5.12	The C-V curve of SRD.	134
5.13	(a) Conventional model of SRD and (b) Fully functional SRD model for CAD simulation.	135
5.14	The schematic circuit of an SRD.	138
5.15	The input waveform and that of the impulse at the diode/transmission line node.	140
5.16	Measured (blue curves) and simulated (red curves) S-parameters of coupled line BPF.	142
5.17	Simulated Power spectrum at filter output.	143
5.18	Power spectrum at SRD output.	144
5.19	Experimental results (a) output power vs input power (b) noise floor vs input power	145
5.20	Output power vs transmission line length at 18th, 19th and 20th harmonics.	145
5.21	The actual circuit board of SRD and BPF.	146
5.22	Schematic of two-stage amplifier.	148
5.23	ADS simulation insertion loss, return loss of bias tee circuit.	149
5.24	Photograph of the two stage 6.8GHZ amplifier prototype.	151
5.25	Measured S–parameters (magnitude) of the two stage amplifier CFY-67 against frequency at a drain bias voltage Vds =3.5V.	153
5.26	Measured output powers, efficiencies and gain of 90MHz amplifier (a) Input power vs output power and efficiency plot (b) Input power vs gain plot.	155
5.27	Measured output powers, efficiencies and gain of 360 MHz amplifier (a) Input power vs output power and efficiency plot (b) Input power vs gain plot.	156
5.28	Power spectrum of 6.834GHz output of Rb synthesizer.	158
5.29	Power spectrum of second harmonic distortion level.	159
5.30	Power spectrum of sub harmonic signal to the SRD.	162

List of Figures   xxi

5.31   Measured output phase noise spectrum of the 10 MHz OCXO. 163
5.32   Measured output phase noise spectrum of R&S instrument (black curve) and clock synthesizer (red curve) at 6.834 GHz.The green plot shows the phase noise spectrum of OCXO at 10 MHz. 163
6.1   (a) TE_{011} cavity model with dielectric loading (b) Magnetic field in the meridian plane, mode TE_{011}. 173
6.2   Variation of cavity length versus resonance frequency with three different dielectric constant. 174
6.3   The cavity dimensions (a) without dielectric loading and (b) with dielectric loading. Dimensions not to be scaled. 175
6.4   Magnetic field in the meridian plane-mode TE_{111}. 176
6.5   Realized prototype microwave cavity with tuning plunger and Teflon ring. 178
6.6   Measured data of $|S_{11}|$ using a 6.834 GHz cavity. 179
6.7   Measured Q circle data of the input impedance of the 6.834 GHz cavity, plotted on the Smith Chart to demonstrate the tuning effect of the variable tunner. The resonant frequency f_L is indicated by 1. 179
6.8   Measured phase of S_{11} for a 6.834GHz cavity. 180
6.9 S_{21} transmission as simulated by HFFS with Q = 226 at f = 6.834 GHz. 181
6.10   Experimental determination of Q-factor, with Q = 194 at f = 6.834 GHz. 182
6.11   Cavity assembly with photodiode detector and absorption cell. 183
6.12   Experimental measurement of TC for TE_{111} cavity. . . . 185
7.1   Experimental setup for physics package discriminator signal characterization. 192
7.2   Discriminator signal at the output of lock-in amplifier shows the error signal that is derived from the absorption signal by the lock-in amplifier at different RF frequencies. 193

7.3	Absorption-resonance signals.	194
7.4	Block diagram of the analog lock-in amplifier.	195
7.5	The circuit details of TIA.	198
7.6	(a) TIA frequency response for $Cp = 0 - 100\,\text{pF}$ (b) simulation of the TIA output noise spectrum with $Cp = 0 - 100\,\text{pF}$.	199
7.7	Delyiannis–Friend filter schematic.	200
7.8	Schematic of phase shifter circuit.	201
7.9	Simulated results of Phase shift vs resistance.	203
7.10	Basic block diagram of mixer circuit.	204
7.11	Mixer output when RF (a) above (b) below (c) equal to the resonance frequency.	205
7.12	The error signal after the LPF (a) error voltage is $+4.9$ V when $f_{RF} = f_{res} - 300$ Hz (b) error voltage is 0 V when $f_{RF} = f_{res}$ (c) error voltage is -4.4 V when $f_{RF} = f_{res} + 300$.	206
7.13	Integrated test setup for SNR measurement.	207
7.14	Measured SNR of detected signal: 70.6 dB.	208
7.15	Measured stability of the clock.	209
8.1	The Rb bulb assembly Unit. (a) Simulated view (b) Experimental view.	213
8.2	Assembly views of Microwave cavity. (a) Simulated view (b) Experimental view.	214
8.3	The plot of demagnetization factor vs L/R.	219
8.4	Plot of the magnetic field variation with DC current.	220
8.5	Schematic of three layer magnetic shield assembly for the Physics package.	222
9.1(a)	The open loop experimental set-up of the Rb clock.	224
9.1(b)	Conceptual block schematic of Servo section.	225
9.2	(a) double resonance absorption spectrum and (b) slope of the discriminator error signal.	226
9.3	The resonance 2nd harmonic signal vs. temperature plot.	227
9.4	Plot of 2nd harmonic signal versus the RF power.	228
9.5	The variation of clock frequency with RF power.	229

9.6   Plot of (a) Transition width vs. modulation index
      and (b) 2nd Harmonic signal vs. modulation index. . . . 229
9.7   The optimized value of modulation index. 230
9.8   Absorption cell temperature vs Rb resonance
      frequency. 232
9.9   Open view of tested ETM of the Rb atomic
      frequency standard. 235
9.10 Allan deviation plot of unlocked and locked OCXO. . . . 236
10.1 Diagram of Rb bulb and cell filling ultra high
      vacuum system. 240
10.2 Rb bulb/Cell filling ultra-high vacuum system. 241
10.3 Drawing of the Rb absorption cell. 242
10.4 Drawing of Rb bulbs. 242
10.5 Block diagram of lamp temperature controller. 255
10.6 Integrate and Dump servo section. 257
10.7 RF section block diagram using frequency multiplier
      topology. 258
10.8 All digital servo system. 259
10.9 RF section block diagram using PLL topology. 259
10.10 Schematic of RF generation section. 260
10.11 Simulated response of 10 MHz × 3 multiplier. 262
10.12 Simulated response of 30 MHz × 3 multiplier. 263
10.13 Simulated response of SRD multiplier. 265
10.14 RF section using PLL approach. 266
10.15 Loop-gap microwave cavity. 268

Glossary and Acronyms

ADC	Analog to Digital Converter
AD	Allan Deviation
ADS	Advanced Design System
AFS	Atomic frequency standard or "atomic clock"
BTC	Baseplate Temperature Controller
C-Field	D.C magnetic bias field applied to Rb Physics Package
CPT	Coherent Population Trapping
CSAC	Chip Scale Atomic Clock
DAC	Digital to Analog Converter
DDS	Direct Digital Synthesizer
DSP	Digital Signal Processing
DVM	Design Verification Model
EMI	Electromagnetic Interference
ETM	Engineering Thermal Model
GLONASS	Global Russian Navigation Satellite System
GPS	Global Positioning System
HEMT	High Electron Mobility Transistor
HFFS	High Frequency Field Simulation
IRNSS	Indian Regional Navigation Satellite System
ISRO	Indian Space Research Organization
LIA	Lock-in Amplifier
LS	Light Shift
MCS	Master Control Station
MS	Monitoring Station
NBS	National Bureau of Standard

NPL	National Physical Laboratory UK
NPLI	National Physical Laboratory India
NIST	National Institute of Standards and Technology
OCXO	Oven Controlled Crystal Oscillator
PSC	Pressure Shift Coefficient
PSD	Phase Sensitive Detector
QTP	Qualification Test Procedure
R & QA	Research and Quality Assurance
Rb	Chemical symbol for Rubidium
^{85}Rb	Most (72%) abundant Rb isotope, used for hyperfine filter
^{87}Rb	Other Rb stable isotope 28% abundant, used as hyperfine frequency reference
RF	Radio Frequency
RFS	Rubidium frequency standard
S/N	Signal-to-Noise Ratio
SSB	Single Sideband
SRD	Step Recovery Diode
TC	Temperature Coefficient
TFR	Thin Film Resistor
TIA	Trans-impedance Amplifier
VCXO	Voltage Controlled Crystal Oscillator
VCSEL	Vertical Cavity Surface Emission Laser

Introduction

> The navigation satellite systems have on-board Rb atomic clocks, which are light weight, small sized, excellent in short-term frequency stability and relatively inexpensive. All over the world, more and more people find apps based services, particularly, hiring Taxis and telecom as indispensable part of their lives. The curiosity is bound to arise to know how Rb atomic clocks make it possible, that too when more exotic and state-of-the-art clocks are available. To set it at rest, we describe in this book, the Rb atomic clocks in simplistic way, and also provide the recipe to produce them. We include a brief historical background of Time keeping since time immemorial.

The Rubidium (Rb) atomic clocks are the workhorse of the satellite navigation systems of which GPS is now a household name. With a few press of buttons, navigators and vehicle drivers all over the world reach their destinations effortlessly with high precision. People are curious to know what makes it possible. To set at rest their curiosity, the need is felt that they are told in simple terms, that no satellite navigation system is thinkable or possible without Rb atomic clocks. This book is all about that. Why choose Rb atomic clocks when far better and exotic atomic clocks are available. The Rb atomic clocks are widely used in satellite navigation systems as they are slim, low in weight, easy to build and also not that expensive. The working of the majority of the atomic clocks is based on transferring the stability of the atomic transitions, generally, between the ground state hyperfine levels, to a VCXO/OCXO. The VCXO/OCXO has the excellent short-term frequency stability, but tends to age with

1

time and that leads to drift in its frequency. This aging or drift in frequency is taken care of, by locking its frequency to the atomic transition frequency. This is how an atomic clock is conceptualized and developed. It is as simple as that.

0.1. Satellite navigation systems — global and regional

We now describe briefly a satellite navigation systems, highlighting the critical role of the on-board Rb atomic clocks. One of the prime satellite navigation systems is GPS, with more than 24 satellites, each carrying on-board three to four atomic clocks. Many of these atomic clocks are Rb atomic clocks. Other bulky atomic clocks, like Caesium(Cs) atomic beam clocks and Hydrogen(H) Masers are also used, mainly for probing their technical acceptability for the space worthiness. GPS, one of the most popular satellite navigation systems, is widely used all over the world. Over the period of time, other global satellite navigation systems were developed by several countries and side by side, some countries also developed Regional Satellite Navigation Systems (RSNS) to meet their regional requirements. As a result, we have now two types of satellite navigation systems: global and regional. The Global Navigation Satellite Systems (GNSS) provide world wide coverage. The others, known as Regional Navigation Satellite Systems (RNSS) provide coverage just to one region, which may be a single country or a sub-continent. Let's briefly describe the satellite navigation systems which are operational or under development.

(A) Global Navigation Satellite Systems (GNSS)

(i) *Global Positioning System (GPS)*

The NAVSTAR GPS system developed by U.S. Department of Defence, consists of the minimum 24 satellites orbiting in synchronous orbits at the altitude of approximately 20000 km and each satellite completes two rounds around the Earth in a day. Four or more satellites of GPS can be accessed from anywhere on or near the

Earth surface simultaneously. The system provides critical capabilities to military, civil and commercial users worldwide and is freely accessible to anyone with a GPS receiver. However, there were earlier accessibility restrictions introduced by US military, which stood withdrawn since 2002. Now anyone can utilise the full potential of GPS for free.

(ii) *Russian Global Navigation Satellite System (GLONASS)*

Soviet Union initiated the development of GLONASS in 1976 operated by the Russian Aerospace Defence Forces. The first GLONASS satellite was launched in 1982 and the system became fully operational in 1993 and the latest generation is acronym GLONASS-M. Currently, GLONASS has 24 satellites in the constellation. Similar to GPS, the GLONASS has variable number of satellites visible at a point, depending on the location. When a minimum of four satellites are in the view, a GLONASS receiver may determine the position in three dimensions and also synchronize time. GLONASS is the only other navigational system in operation with global coverage and of comparable precision of GPS.

(iii) *Galileo*

Galileo is another global navigation satellite system launched by European Union and European Space Agency. It is intended primarily for civilian uses. Galileo provides a high-precision positioning system for European nations. However, it is a global NSS like GPS and GLONASS. The 30-satellite navigation system will be operational in 2019. This navigation system will have both free and paid components. The use of the basic services are free and open to everyone, while the high-precision capabilities will be available for commercial and military uses with payment of some fee.

(iv) *Compass (under development)*

The Compass global navigation satellite system is being developed by China. It is an upgrade of its second generation regional BeiDou Satellite Navigation System (BDS), also known as BeiDou-2. It is likely to be completed in 2020, consisting of 35 satellites. Basically, it is meant to be an alternative to GPS, considering various geopolitical possibilities, that may arise in the future.

(B) Regional Navigation Satellite Systems RNSS

(i) *Quasi-Zenith Satellite System (QZSS)*

The Quasi-Zenith Satellite System (QZSS) is a three-satellite regional time transfer system and Satellite Based Augmentation System for the Global Positioning System. It is proposed for Japan and Australia. QZSS will have mobile applications and provide telecommunication based services and the positioning information. Its three satellites, each 120° apart, are in highly-inclined, slightly elliptical, geosynchronous orbits. Because of this, they do not remain in the same place in the sky. Their ground traces are asymmetrical figure-8 patterns, created to ensure that one is almost directly over Japan at all times.

(ii) *BeiDou Navigation Satellite System (BDS)*

The BeiDou Navigation Satellite System (BDS) consists of two separate satellite constellations. The first is a limited test system, known as BeiDou-1. It has been in operation since the year 2000. The BeiDou-1 consists of three satellites and offers limited coverage, navigation services and applications mainly for users in China and neighbouring regions. As mentioned previously, the second generation is a full-scale global navigation system that is currently under construction and will be known as Compass, or BeiDou-2.

(iii) *Indian Regional Navigational Satellite System (IRNSS)*

The Indian Space Research Organization is developing the Indian Regional Navigational Satellite System (IRNSS), a regional satellite navigation system. It is a cluster of 7 satellites, covering 1500 km in India and its neighbourhood. Each satellite contains 3 Rb atomic clocks for providing positional accuracy of 20 metres. After completion, it will be under the control of the Indian government. IRNSS will provide standard navigation services for civilian use and an encrypted restricted service for authorized users (military). ISRO Space Application Centre SAC, Ahmedabad India in collaboration with CSIR

NPL New Delhi India is given the responsibility to design, develop and deliver payloads for IRNSS.

The satellites in the above mentioned GNSS/RNSS, have onboard atomic clocks, controlled by a master control centre and tracked by several monitoring stations. The master control centre provides necessary correction to the course of these satellites and keep the onboard atomic clocks in sync. The GPS is based on trilateration (a method of determining positions using the points of intersection of three overlapping circles or spheres) i.e., to know the position of any person or an object accurately, the minimum three satellites should be simultaneously trackable at that point and to know the time/altitude, one additional satellite should be monitored at that point. The master control centre can receive and transmit signals to the satellites and the onboard payloads, for correcting the path of satellites and synchronizing the onboard Rb atomic clocks. However, the monitoring centres and the users can receive only the coded information 24/7. The coded signals carry the information on the distance and time, from each satellite. By simultaneously tracking three to four satellites, the user can get their position and time accurately. This information can also be used to track any vehicle or moving object. The mobile service providers and mobiles Apps, process the satellite navigation system signals, to know and track the positions of users. Presently, among the satellite navigation systems, GPS has become the essential part of everyone's life. The role of the onboard Rb atomic clocks is extremely critical, as an inaccuracy of even 1/100th of a Second may lead to a distance uncertainty of 3000 km. One can very well appreciate how accurate the Rb atomic clocks are to be, for the distance accuracy of a few meters. With the necessary path and ionospheric corrections and in the differential GPS mode, an accuracy of a few centimetres is also possible. Needless to mention, the atomic clocks particularly, the Rb atomic clocks, onboard and in the master control centres, play the most critical role and without them, the navigation satellite systems can not work. In the nutshell, an accurate timing maintained, by the onboard Rb atomic clocks is the key to measuring precise distance to the satellites. The navigation satellite systems are highly accurate with the onboard atomic clocks. The receiver clocks need not be too accurate. Because, the trilateration method, with the visibility of minimum a three satellites at that place, can remove the errors and provide precise Time and positional accuracy of a few metres. The following photos, A and B, give the idea of the workings of GNSS. With the above details on the importance and indispensability of the Rb atomic clocks, in any navigation satellite system, the readers are naturally curious to know about the Rb atomic clocks. This heightened curiosity is to be kept alive a little longer. Before we unfold the details of the Rb atomic clock, we will take you down the interesting and captivating memory lane and describe the history of time keeping since immemorial.

The archaeological evidence indicates, that since prehistoric times, human beings were devising progressively better means of time keeping. In the earliest stage, this involved observation of the

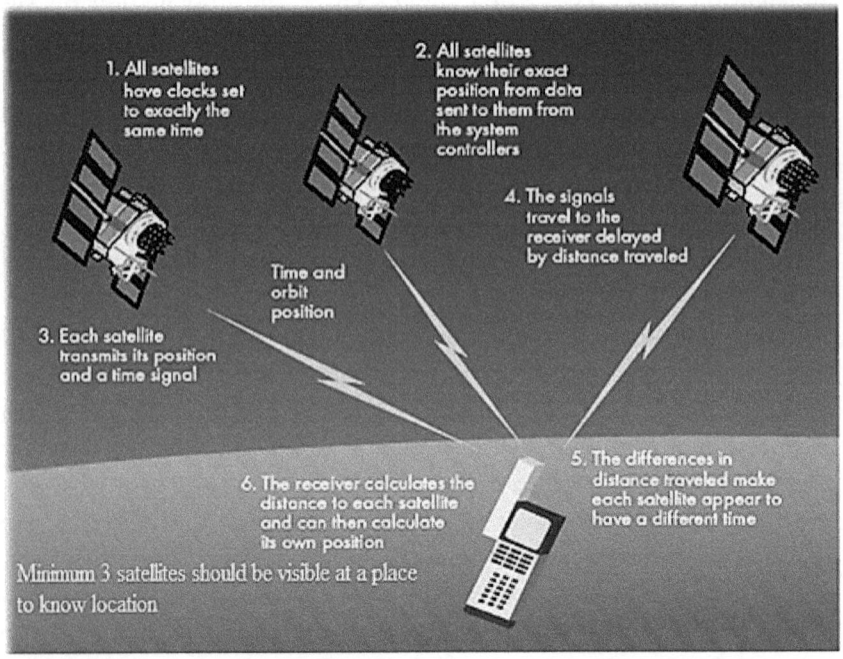

Photo A: Working principle of a satellite navigation system.

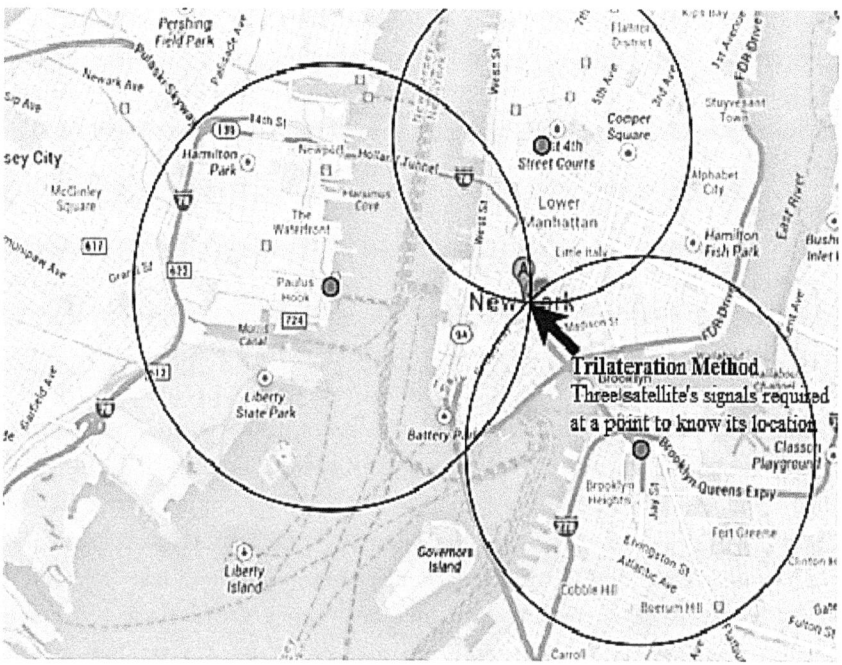

Photo B: Precise position determination of any object requires simultaneous monitoring/visibility of at least three satellites — Trilateration method.

apparent motion of the sun, but finer subdivision of the day later involved devices such as water clocks, hourglasses, and calibrated candles. Over a period of time, different types Time measurement methods and Time standards were used, for example, astronomical cycles in sundials, pendulums in mechanical clocks and quartz crystals in the electrical clocks [1–2]. However, with the invention of the two-pendulum clock in 1921 by William Hamilton Shortt, the upper limit of such mechanical clocks was reached. The modern era of timekeeping began with the development of the quartz crystal oscillator. In a patent application in the year 1918, Alexander M. Nicholson disclosed a piezoelectric crystal as the controlling element in a vacuum tube oscillator. Joseph W. Horton and Warren A. Marrison subsequently developed the first clock, controlled by a quartz crystal in 1927. Since the introduction of the quartz oscillator, the performance of the frequency standards advanced in terms of frequency stability and accuracy by several orders of magnitude. The industries, science and technology started depending on the accuracy and stability of these crystal oscillator untill the advent of atomic clocks. In the year, 1949, the atomic timekeeping era began with the construction of the first Ammonia molecular/atomic clock. The molecular/atomic clock, based on a resonance in the Ammonia molecule, was constructed under a project led by Harold Lyons [3–5] at the former National Bureau of Standards, presently known as NIST, USA. The Ammonia Frequency and Time standard was subsequently, superseded by the Caesium-beam frequency standard that forms the current basis for defining the unit "Second". The atomic clocks progressed rapidly in accuracy and stability through the years 1950s and 1960s [6]. By 1968, in the International System of Units (SI), the unit of Time "Second" was redefined to be exactly 9,192,631,770 cycles of oscillation corresponding to a transition between the two hyperfine sublevels of a ground-state ^{133}Cs atom [7–8]. Since then, atomic clocks have been developed, based on the transitions between ground-state hyperfine levels of either H or an alkali atoms, typically Cs or Rb [9–14]. However, the development of new and improved atomic clocks is the ongoing research and development work. The atomic clocks that use other atoms and techniques, for example, trapped-ion cold atoms and optical lattice continue to be the subject of intense research and are projected as future Time and Frequency Standards [15–16]. Out of the many types of atomic clocks, the most widely used, specifically, for satellite navigation, is the vapour-cell based Rb atomic clock. It is relatively simple to develop, as it is based on vapour cell containing a dilute vapour of Rb and an inert buffer gas at a few Torr pressure [9, 17]. The state selection is done using a small glass cell unlike bulky multi-poles magnets in H or Cs standards. The Rb frequency standard is the only one among all atomic frequency standards, that has probably received so much attention for a long period. It is still the subject of intensive research and development. In fact, its characteristics are such, that it is one of the most useful frequency standards when the size, reliability and good short and medium term frequency stability are desired. It finds wide applications in the satellite navigation, tele-communication, stand-alone frequency Standard, reference frequency source in precision instruments.

Because of its wide raging applications, even today there is continuous effort to reduce its size [18] and to improve its frequency characteristics [19–20]. It is pertinent to mention that the Rb atomic clock is extremely useful and light in weight, as the state selection is done using a small Rb glass filter cell, which may be separate or integrated with the Rb lamp and Rb absorption cell. These integrated and separate filter cell configurations are discussed in sections 1.1.3 and 1.1.4 respectively. In the other atomic clocks like H-Maser, Cs beam tube atomic clocks, the state selection and the clock signal detection are done using heavy hex pole magnets. That makes them not suitable for satellite navigation. However, in the ongoing development, the diode Lasers are replacing the heavy magnets in some of these atomic clocks, but for satellite navigation, their reliability is still an issue, so these clocks are not in vogue.

Chapter 1

Overview of Rb Atomic Clocks and the Space Specifications

> We describe the functioning of Rb atomic clock including its two important components. Namely, the Physics and the electronic packages. The energy levels of the Rb isotopes and the optical pumping for population inversion are described. The detailed specifications of space qualified Rb atomic clocks are also discussed.

1.1. Working principle of Rb atomic clock

Rubidium (Rb) is an alkali metal like Lithium, Sodium, Potassium and Caesium (Cs). Rb is a silvery metallic element of the alkali metals, in the first main group of the periodic table, with the atomic number 37. There are two, naturally, occurring stable isotopes of Rubidium, ^{85}Rb and ^{87}Rb, which have 72.2% and 27.8% natural abundance respectively. The metal has a melting point of 39°C. The alkali metals in general, have similar characteristics, as they all have one electron in the outermost electronic orbit, outside an inert core. This outer electron determines the most of the chemical, electronic and spectroscopic properties of these elements. Rb alkali atoms, when excited, radiate in near infrared region mainly with wavelengths 780 and 794.7 nm. Rb and Cs are used in the atomic clocks, because they have a single valence electron and their properties can be studied accurately, using single electron model. Besides, they have the ground state (g.s) hyperfine transition frequency in the microwave region, which can easily be multiplied and synthesized, for getting

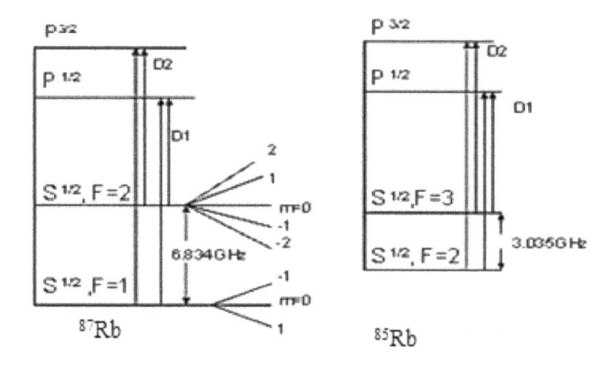

Fig. 1.1   Energy levels of ^{87}Rb and ^{85}Rb.

the clock error signal. The environmental effects on these atoms are also manageable.

The Rb atomic clock works on the principle of phase locking a voltage control crystal oscillator (VCXO) to the frequency of transitions between the ground state (g.s) hyperfine sublevels F = 2, m = 0 and F = 1, m = 0 of ^{87}Rb atom, as shown in the Fig. 1.1. These levels are chosen, as they have second order dependence on the magnetic field. These energy levels are the least affected by the environmental or earth's magnetic field and its fluctuations. To obtain the necessary correction signal for the phase locking VCXO, the optical pumping technique is used for creating the population inversion. The light from ^{87}Rb bulb, after filtering through a ^{85}Rb filter cell is incident on a ^{87}Rb absorption cell. Which is kept inside a microwave cavity, tuned to the ^{87}Rb hyperfine transition frequency, 6.834 GHz. The g.s hyperfine levels F = 2 of ^{87}Rb and F = 3 of ^{85}Rb overlap, as shown in Fig. 1.2. As a result, the filtered light, with almost no transition from F = 2 level to P excited states of ^{87}Rb atoms, induces transitions from only F = 1 g.s hyperfine sublevel to the excited P levels. While relaxing back to the g.s, the atoms have almost equal probability to occupy F = 1 and F = 2 levels. Thus, the population of F = 2 level is increased, while completely depleting F = 1 level after several such cycles. A steady state condition of the population inversion is obtained. The ^{87}Rb absorption cell becomes more transparent to the incident light. In this state of population inversion, the

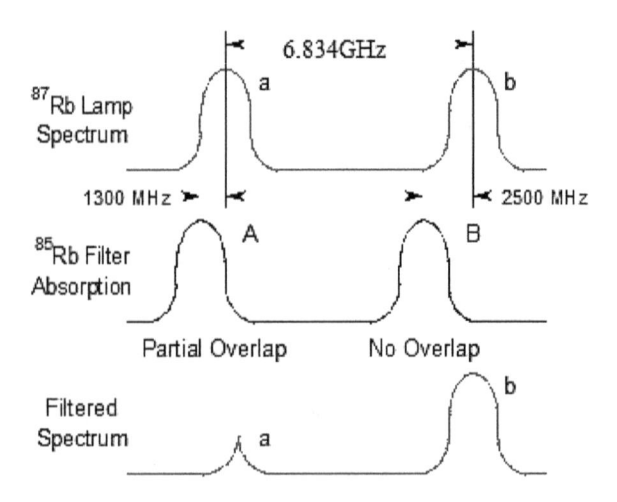

Fig. 1.2 The g.s hyperfine components with the ^{87}Rb radiation line-a filtered by the ^{85}Rb line-A overlap, ^{87}Rb radiation line-b available for the optical pumping.

application of the resonant field microwave field, with its magnetic field in the direction of the light axis, results in a net transition of atoms from F = 2 to F = 1 level. For restoring the steady state, more light is absorbed by ^{87}Rb atoms in the absorption cell. The transmitted light from the absorption cell shows a dip in the intensity of the light signal, monitored by a photo-voltaic cell, Figs. 1.3 and 1.4. If the microwave field is phase modulated at a low frequency then the resonance signal shows a composite nature, containing the first and second harmonics of modulating signal, riding on the DC signal. The resonant signal after phase sensitive detection and passing through a low pass filter, produces a DC correction signal, which is applied to the electronic frequency control of VCXO. In this way the stability of the g.s hyperfine transition frequency is transferred to the VCXO.

The Rb Time and Frequency standards or atomic clocks currently used as the frequency references, are very compact, lightweight, require low-power and are portable as well. These unique features of Rb atomic clocks, make them first choice of the satellite navigation systems. The Rb atomic clock broadly, has two packages. The Rb Physics and the Electronic packages. The heart of the Rb atomic clock is the Physics package. It contains a ^{87}Rb lamp, excited by a RF circuit, a ^{85}Rb filter cell and the absorption cell containing isotope-enriched ^{87}Rb vapour, in the separate filter cell technique. While in an integrated filter cell technique, the filter and the absorption cells are merged in one glass cell. The ground-state hyperfine transition is interrogated by a resonant RF field produced by a tuned microwave cavity, which houses the absorption cell. As mentioned above, the Physics package has two configurations (i) Integrated Filter cell Technique (IFT) (ii) Separate Filter cell Technique (SFT). Both these techniques are discussed later in this section. It is interesting to note, that the IFT based Rb clocks were first to be used in GPS and Galileo, as it saves power by eliminating the heater for the filter cell. In many of the Rb atomic clocks, the diode laser or VCSEL has replaced Rb electrode-less lamps. However, in the satellite navigation systems, the Rb lamps are preferred over diode lasers. Its shot-noise limited frequency stability is better than that of the laser-pumped Rb atomic clocks. It can be emphatically mentioned, that the Rb lamp based clocks remain unchallenged for the global navigation satellite systems and will be used for many years in future. Therefore, in this book, the Rb lamp based Rb atomic clocks are discussed in detail. The ^{87}Rb atoms under normal conditions, have both the ground state hyperfine levels populated almost equally. The optical radiation from the Rb lamp, after filtering of the undesired optical line and incident on the ^{87}Rb absorption cell, produces a population imbalance far in excess of that which exists under thermal equilibrium. This method of population inversion is known as the optical pumping. The energy separation of ground state (g.s) hyperfine levels is approximately 6834 MHz ($\approx$ 28 μeV).When the atoms in the absorption cell are subjected to a resonant microwave RF field, the transitions between g.s hyperfine levels are induced, which tend to reduce the population difference built up by the optical pumping. It reduces the amount of optical radiation transmitted through the absorption cell and incident on the photo detector. The phase or frequency modulation of the resonant microwave RF field, by a small modulation index at some low frequency (137 Hz), modulates the intensity of the light reaching the photo detector. The photo detector output contains the modulating frequency and its second harmonic (274 Hz), riding on a DC signal. The amplitude of the modulated light intensity is proportional to the slope of the microwave resonance absorption line. By synchronously demodulating the amplified output of the photo detector, a dc error voltage, known as the clock signal is obtained. This clock signal is applied to the electronic frequency control of OCXO for correcting its frequency. As the crystal oscillator (OCXO) frequency is multiplied and synthesized, the resonant microwave field is produced. The frequency corrected OCXO is the Rb atomic Time and Frequency Standard.

As mentioned above, the Rb atomic clock has two important components i.e., (i) Physics package (ii) Electronic Package, controlling the RF and allied electronics to stabilize the input and output parameters of the Physics package. We first describe the Rb Physics package, the most critical part of Rb atomic clock, as it determines the overall performance of the clock. It is relevant to mention here that the cost of a Rb Physics package is almost $3/4^{th}$ that of the Rb atomic clock. Its technology is a guarded secret by manufacturers.

1.1.1. Rb physics package

The Core of the Physics package consists of a Rb bulb, integrated or separated filter cell and the microwave cavity. The Physics package also includes a solenoid providing necessary C-field, a triple magnetic shield layer, to reduce effect of environmental fluctuations of the magnetic field and two bifilar (non-magnetic field producing) heaters to control the temperature of the Physics package at the uncertainty of 0.1°C or lesser. As mentioned earlier, in the integrated filter cell technique (IFT), the optical resonance radiation from a Rb lamp is transmitted through a glass absorption cell containing Rb in vapour state and a buffer gas at low pressure. The buffer gases reduce Doppler broadening in the microwave resonance frequency of interest [17]. In the separate filter cell technique (SFT), a filter cell containing ^{85}Rb isotope is interposed between the lamp and the absorption cell.

1.1.2. Integrated filter cell technique (IFT)

The integrated filter cell configuration has a Rb bulb and a tuned microwave cavity, with a Rb absorption cell inside it, Fig. 1.3. In this configuration no separate Rb filter cell is used. In the integrated filter cell configuration, the Rb lamp consists of electrode-less Rb bulb and a lamp exciter. The Rb bulb is made of Pyrex 7070/7740/Schott 8436/GE-180 glass in spherical shape of outer diameter 10 mm and wall thickness of 0.5 to 0.7 mm. The bulb is filled with ^{87}Rb isotope and natural Rb of 99.99% purity in 1:1 ratio, and each in the quantity of 350 µg (approx). The use of the mixture of natural and ^{87}Rb in equal proportions, ensures the minimization of the light shift. The bulb is also filled with Krypton/Xenon gas at 2.0 ± 0.2 Torr of nearly 99.995% purity for the ease of excitation as Krypton/Xenon has low ionization potential. The lamp is excited by a 55–100 MHz, 2–3 watt RF oscillator. It is a Colpitt/Clapp oscillator run on a D.C power supply of rating 20 volt to 24 volt and 0.2 to 0.4 Amp. The light intensity and the mode of operation of the lamp may be controlled by changing the gain and frequency of the oscillator, with the help

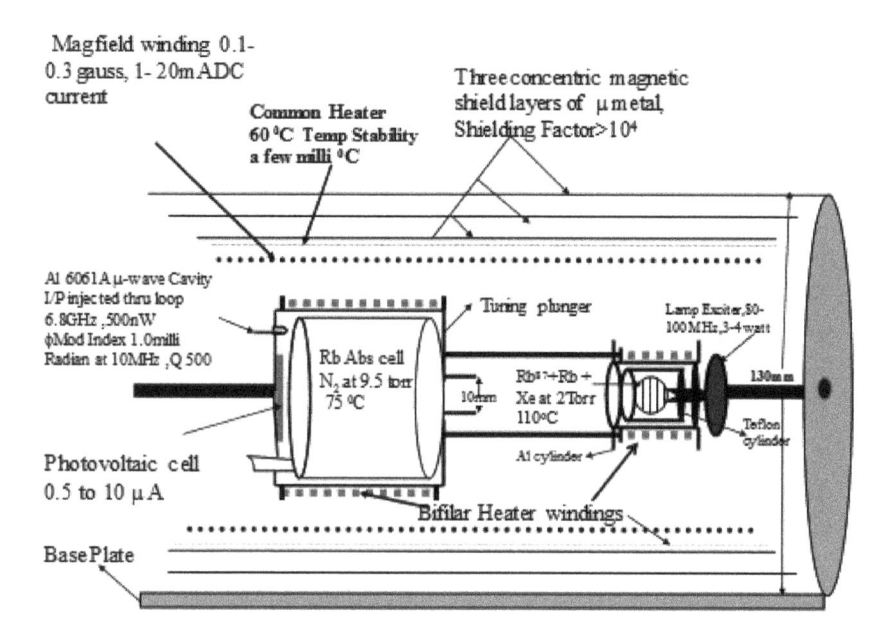

Fig. 1.3 Rb physics package-integrated filter cell.

of a resistor and capacitor respectively. As the Rb lamp is a highly critical component of the Physics package it is necessary to keep provision to control these lamp exciter parameters precisely. In this section, the light shift, a result of the distortion in the lamp spectral profile and intensity fluctuations, and also the buffer gas broadening of the atomic resonance line are discussed. The light shift severely affects the performance of the Rb atomic clock. The lamp is self r-f heated but it requires additional temperature controller for the stabilization of its intensity and the operation in the desired mode. When the temperature is around 110°C , the lamp operates in a ring-mode [20, 21] i.e., the Rb light appears to be emitted from a narrow ring, close to the surface of the bulb and the purple colour Rb light has very narrow line width and there is no self-reversal. The self-reversal is a phenomenon in which the spectral line shows a depression in the centre of the line profile. When the temperature is increased, the intensity decreases and the colour of the light becomes deep purple or reddish. The spectral lines become broad and highly self-reversed [12]. In view of the above mentioned modes at different temperatures of the lamp, it is desirable that for the ring mode, the lamp's temperature should be stabilized at 110°C with the stability ±0.1 Celsius. We discuss the reasons of the transition from the ring to the red mode. Various groups have studied the phenomenon. Camparo [20] and Shah R.S. [22] have described the mode transition due to the radiation trapping but their explanations are not fully satisfactory. Camparo has made the ad-hoc assumption that P-levels of Rb atoms become metastable and this gives rise to the radiation trapping.

To describe the Rb lamp mode transition, we should consider the coherent population trapping (CPT) [12] of Rb atoms in the g.s state hyperfine levels. To elaborate on it, we consider Dicke's [17] seminal work on the presence of partial coherence in the spontaneous emission process. The partial coherence in the spontaneous emission leads to the confinement of unexcited Rb atoms to the g.s hyperfine sublevels, due to the destructive interference between the two radiation lines, originating out of the transitions from the g.s hyperfine levels $5S_{1/2}$ F = 2 and F = 1, to $5P_{1/2}$ and $5P_{3/2}$ in the Rb lamp. This trapping of some of the atoms in the g.s hyperfine levels is equivalent to the situation, as if there is radiation trapping in the excited state, as inferred by Camparo. However, Camparo has assumed that the excited states behave as metastable states. We are of the opinion, as mentioned above, that this radiation trapping and the mode change in the Rb lamp may be the result of the trapping of the atoms in the g.s hyperfine levels, due to the partial coherence in the spontaneous emission. Besides, in the integrated filter cell technique, the filtering of the undesired light may not be that efficient. Therefore, the presence of the coherent part in the undesired radiation line as well as in the desired radiation line, may coherently trap some of the ^{87}Rb atoms in the ground state. The trapping of the atoms due to CPT is, physically, manifested in the loss of optical pumping efficiency in the Rb absorption cell. Based on the above analysis, it may be established qualitatively that the radiation trapping phenomenon is, basically, produced due to the CPT of the atoms in the ground states, both in the Rb bulb and the absorption cell. The quantitative analysis of the radiation trapping due to CPT requires separate detailed experimentation. However, based on Dicke's work and the above discussions, it may be explained that the radiation trapping results due to CPT of the Rb atoms.

In the Rb Physics package, the residual temperature coefficient generally is of the order of 10^{-10} °C^{-1}. Therefore, for achieving the frequency stability of the order of 10^{-13} or better, the lamp and cell temperatures should have stability of the order of the fractions of a Celsius. To control the temperature of the Rb lamp and the integrated Rb absorption cell, separate heaters are provided. These heaters, with proportional current control, have bifilar windings for eliminating the magnetic field, produced by the current passing through them. The glass-encapsulated thermistor is placed beneath these heaters for sensing the temperature and providing the necessary input to the temperature controller circuits. In addition to these two heaters, one common heater for maintaining the temperature a few degrees above the ambient temperature is used. This double oven approach ensures temperature stability of a small fraction of a Celsius. The heaters are powered by 15 and 20 volt D.C supplies respectively.

In the integrated filter cell configuration, the absorption cell contains the natural Rb approx. 1mg and Nitrogen as a buffer gas at 9.5 Torr. The cell is made of Pyrex glass/Schott 8436/GE180 so that the quality factor of the microwave cavity remains, practically, unaffected. The Nitrogen gas also provides quenching of the scattered and fluorescent radiation. This property helps in preventing radiation trapping, which tends to affect the efficient optical pumping of the Rb atoms [23].

1.1.3. *Separate filter cell technique (SFT)*

In separate filter cell technique (SFT) there are three main components, ^{87}Rb bulb, a ^{85}Rb filter cell and ^{87}Rb absorption cell, Fig. 1.4. The basic purpose of introducing a separate ^{85}Rb filter cell is to have

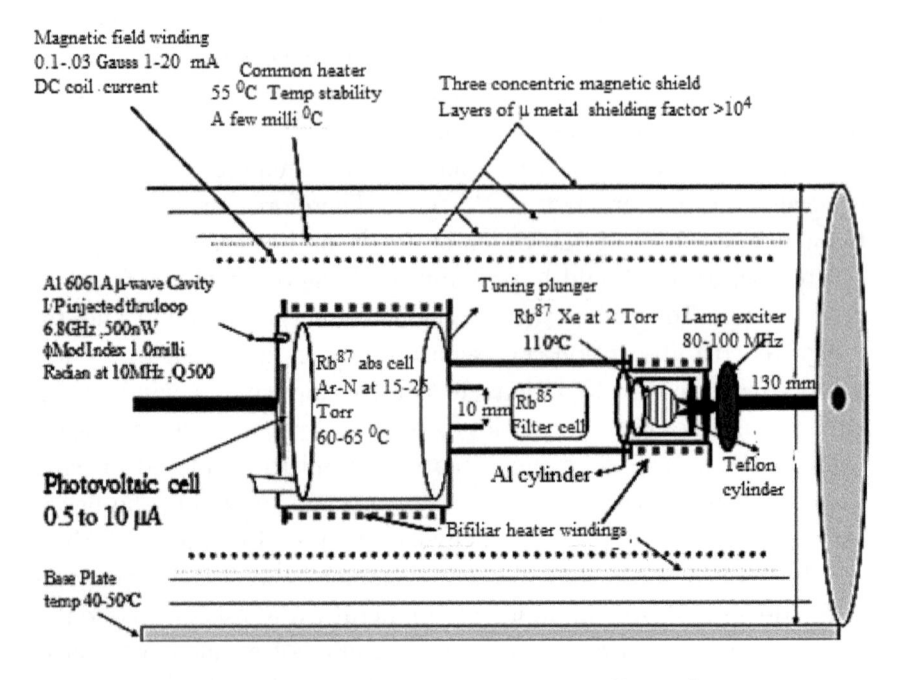

Fig. 1.4 Rb physics package-separate filter cell.

efficient filtering of undesired radiation. In the case of stable Rb isotopes i.e., ^{85}Rb and ^{87}Rb, the ground state hyperfine level F = 2 of ^{87}Rb more or less matches with F = 3 level of ^{85}Rb atoms in filter cell Fig. 1.2. In the SFT, the lamp contains ^{87}Rb isotope with Kr/Xe high purity gas. The light passing through this filter cell does not allow undesired radiation to be incident on ^{87}Rb absorption cell. Only the desired radiation, required for the efficient optical pumping of ^{87}Rb atoms, reaches the absorption cell. This configuration may completely eliminate light shift, which is not possible in IFT. We discuss in the section, the design parameter analysis, the light shift and why it should be eliminated or minimised, for Rb atomic clock, to have an excellent frequency stability.

1.1.4. *Design parameters analysis*

In this section, we discuss various parameters, which affect the frequency stability and the accuracy of Rb atomic clock. The quality of the atomic clock resonance signal is also very crucial. The main factors affecting the performance of the Rb atomic clock are the frequency shifts, which arise due the following reasons:

1. Pressure shift.
2. Light shift.
3. RF power shift.
4. Cavity pulling.
5. Effects of modulation and demodulation.
6. Space and radiation effect.

The effects and causes of these frequency shifts are discussed in detail so that during the development of Rb atomic clocks, these aspects are effectively dealt with.

1. Pressure shift

In Rb atomic clocks, the Rb absorption cells are filled with the buffer gas for preventing/or delaying the relaxation of the Rb atoms on the wall of the cell, so that the narrow linewidth may be obtained.

The collisions of the atoms with the wall lead to the broadening of the line. The presence of the buffer gas delays the collisions with the wall and due to the Dicke's effect, the Doppler broadening is also reduced. However, the presence of the buffer gas in the Rb absorption cell produces a frequency shift. The buffer gas frequency shift is a quadratic function of the temperature. While sealing the cell, the buffer gas pressure changes due to the sealing temperature. This slight uncertainty in the buffer gas pressure leads to frequency shifts. The filtering at 75°C creates a negative temperature coefficient while the Nitrogen buffer gas creates a positive temperature coefficient in IFT. With Nitrogen at 9.5 Torr, these two coefficients more or less cancel each other.

2. Light shift

The light shift, also known as AC Stark shift, is caused by the light intensity and its spectral profile, which affects the hyperfine transition frequency. The optical pumping light and its spectral profile, randomly affect the energy levels and thus shift the hyperfine transition frequency. It is estimated that a one per cent change in the Rb lamp intensity may result in transition frequency offset of the order of 10^{-12}. The implication of this shift is that, it results in the positional uncertainty of several metres. Therefore, it is very important to eliminate the light shift for achieving good frequency stability in the Rb atomic clocks and the positional accuracy in the satellite navigation. In IFT, the filtering of the undesired light takes place on the front side of the absorption cell, containing natural Rb. In this case, the light shift can not be completely eliminated. Whereas in SFT, due to the presence of a separate filter cell, the light shift may be completely eliminated. In both the configurations, the choice of the buffer gases and the operating temperatures of the lamp and the absorption cell are important for dealing with the light shift problem. These parameters set the consistency of light intensity passing through the Rb absorption cell. The processing of glass lamp and absorption cell is also crucial in ensuring good working of the lamp and the stabilization of light intensity, so that the light shift is low or null. There are many theoretical explanations put forth to define the reasons behind light shift. In one of the theoretical explanations, the light shift or AC Stark shift is the result of real photons induced virtual transitions in the Rb atoms that shift the energy levels. The interacting radiation light gives self-energy to the desired ground state hyperfine energy levels. One energy level gets positive while other gets negative self-energy, depending on its spin state. This results in an increase in the energy gap between the two levels. Thus, the possibility of real and virtual transitions arises. The higher intensity leads to positive (+ive) frequency shift and vice versa. This qualitative theoretical explanation is widely used to define the cause of the light shift.

3. RF power shift

The nonhomogeneous broadening of the resonance line in the direction of the light beam, due to the optical pumping, leads to the RF power dependence on the resonance frequency and resulting in the RF power frequency shift. The RF power shift is related to the light shift. The RF power frequency shift is observed more in the integrated cell technique, as the filtering also takes place inside the absorption cell itself, with the filtering efficiency varying with the cell length. The RF power shift may be reduced by stabilizing the RF power level. The instability of the RF power should be less than 0.1 dB for obtaining clock frequency stability of the order of 10^{-13} or better. It is observed that RF shift is responsible for affecting the long-term frequency stability of the Rb atomic clock.

4. Cavity pulling

The cavity pulling results from the difference in the frequencies of the tuned cavity and the resonance. It is more pronounced for the active frequency standards. In the case of Rb atomic clocks, it has the limited effect. It is documented, that a 100 kHz detuning gives rise to fractional frequency shift of the order of 10^{-12} for the clock. It is very small relative to other frequency shifts. However, the aim should be to tune the cavity as close to the resonance frequency as possible.

5. Effects of modulation and demodulation

In the Rb atomic clock, the phase or frequency modulation of the microwave signal is used for obtaining the resonance or clock signal. The modulation frequency may be of the order of the resonance linewidth. The demodulation of the signal in the phase sensitive detector requires that the voltage to phase characteristics is linear, for the sine wave modulation. The results are similar for the square wave modulation. However, the basic difference between the two is that, in the case of square wave modulation only two points of voltage to phase curve are used and the linearity is not an important requirement.

6. Space and the radiation effects

The radiation is a major consideration in the design of the space Rb atomic clock. In outer space, the Rb atomic clock is continuously exposed to the energetic particles and secondary radiation. Not only, the VCXO is affected by the total dose and bursts, but the servo electronics in the Rb atomic clock is also subjected to transitory loss of the frequency lock due to the radiation. Some key components in the Rb atomic clock are radiation hardened, to significantly reduce the impact of the radiation. Heavy radiation shielding may also be added to minimize the impact of the radiation. In spite of all this, during the periods of intense solar activity, the Rb atomic clocks have clearly exhibited discernible sensitivity to the radiation. The spaceborne Rb atomic clocks must maintain a high level of accuracy and frequency stability for several years, under the harsh environment of the space. The mechanical design must be such that the Rb atomic clock withstands the enormous shocks and vibrations at the time of launching. The extreme temperature that may be encountered in the space, requires the thermal and mechanical designs to ensure that the Rb atomic clock maintains an excellent frequency stability. This requirement is expressed in terms of the Temperature Coefficient (TC) of the Rb atomic clock. Typically, the magnitude of the change in the output frequency of a Rb atomic clock should be of the order of $5 \times 10^{-13}/°C$, for the temperature variations between

10°C and +50°C. This performance should be achievable in space, under high vacuum, of the order of 10^{-5} Torr. The stringent requirement for the space clocks is also to withstand the electromagnetic interference (EMI). EMI gaskets, feed through filter capacitors, filter boxes and semi-rigid shielded coaxial cables are used to reduce the EMI effects. Another important space requirement is the magnetic sensitivity. Typically, the magnetic coefficient should not be greater than 10^{-12}/Gauss.

1.2. Microwave cavity

In the Rb atomic clocks as mentioned earlier, the g.s hyperfine transitions of ^{87}Rb, in an absorption cell, are used for generating the clock signal. The Rb absorption cell is placed inside a resonant microwave cavity. The microwave cavity is made of aluminium alloy (Al6061) and is generally operated in TE_{111} mode. The microwave cavity is tuned at the frequency 6.834687500 GHz, which is the transitions frequency of ^{87}Rb atoms between the g.s hyperfine levels, F = 2, $m_F = 0, -F = 1, m_F = 0$, Fig. 1.1. The experimental microwave cavity dimensions for TE_{111} mode are; length = 40.0 mm (including tuning length of 10 mm), inner diameter = 27.00 mm and wall thickness = 7 mm. These dimensions can be reduced, considerably, using dielectric material inside the cavity. The microwave feed is through a loop of copper wire of 1.00 mm dia, placed on the rear side of the cavity. For the efficient microwave power injection, the loop is $\lambda/4$ from the inner wall of the cylindrical surface. The Rb absorption cell and the array of the photo-voltaic cells are also mounted inside the microwave cavity. The ground plane of the array of the photo-voltaic cells, acts as the termination of the microwave cavity. On the other end of the microwave cavity, is a tuning plunger, with suitable fine threads for tuning the cavity precisely. While developing Rb atomic clock for the first time, the tuneable plunger is very handy in tuning the cavity exactly to the ^{87}Rb atom ground state hyperfine resonant frequency. However, in the commercial Rb atomic clocks, the length of the cavity is fixed and the tuning plunger is not required. The microwave signal has its magnetic field in the direction of the optical axis. This meets the requirement for exciting the clock transitions between F = 2, $m_F = 0$–F = 1, $m_F = 0$. The clock transitions have second order dependence on the D.C magnetic field, which is applied to remove degeneracy and for defining the quantization axis. The D.C magnetic field can also change the atomic transition frequency. It provides, the clock manufacturers, leverage to match the Rb atomic clock's Second's duration with SI unit of Second. The loaded Q of the microwave cavity is more than 500. The cavity is provided with the bifilar heater winding (no production of any magnetic field) for maintaining the temperature of the microwave cavity and IFT absorption cell at 75°C with the temperature stability ±0.1°C. At 75°C, light shift is nearly zero.

1.3. Effect of photodiode inside/outside microwave cavity and its characteristics

To detect the hyperfine transitions and to obtain the clock signal, arrays of photovoltaic cells/photodiodes are mounted on a PCB. The back side of the PCB is a ground plane. The properties of these photovoltaic cells are very important in deciding the quality of the clock signal. These photodiodes should have very high efficiency near the infrared region, particularly, for the wavelengths 780 and 794 nm. The dark current of these photodiodes should be the minimum for good S/N. The space qualified Rb clocks require photodiodes to be radiation hardened to withstand possible lattice defects, neutron bombardment and radiation induced recombination of the charge carriers. The noise characteristics of the photodiodes are crucial. The studies show that in general, the following noise processes affect their response and the efficiency.

(i) *Photon Noise:* The number of photons detected by photovoltaic cell show statistical fluctuations and depend on the incident light and its frequency. In Rb atomic clock, the optical radiation is in the near infrared region i.e., having longer wavelength. To keep the photon noise low, it is necessary to strictly control the temperature and the intensity of radiation emitted by Rb lamp.

(ii) *Johnson Noise:* The Johnson noise is also known as thermal noise, and it arises due to the thermal fluctuations. To control it, the temperature should be maintained at a constant level.

(iii) *Generation–Combination Noise:* The photo voltaic cell is, basically, a semiconductor diode. The fluctuations in the number of thermally generated electron–hole carriers give rise to generation–combination noise. The life time of these carriers also affects this noise.

There are two other kinds of noise processes namely, modulation and contact noise. Both are similar in nature, with a difference that former is generated at the surface and the latter is found at the metal-semiconductor contacts. These noise processes arise due to dislocation or imperfection in the semiconductors. It is important for the Rb atomic clock, that the photodiodes have the minimum of these noises. The selection of the photodiode should be done very carefully. The photo-detectors are placed either inside or outside the microwave cavity. The simulation results, for the photodiodes with the PCB inside the microwave, show that the PCB backside ground plane acts as the microwave cavity termination. The microwave leakage is reduced, as there is no requirement of a hole in the microwave cavity, on the detector side. This may slightly increase the cavity Q. The photo-detectors cover the cross-sectional area of the absorption cell. In case, the photo-detector is placed outside, a hole in the rear wall of the microwave cavity is required. This results in the leakage of microwave power. Besides, to collect the light transmitting through the absorption cell, a lens is placed in front of the photo-detector. This arrangement does not capture full light intensity, interacting with the Rb atoms. It is observed from the above discussion and simulation, that the former arrangement is better for a higher clock signal.

The detected light signal in general, is of the order of 130–170 millivolt at 75°C. At resonance, the dip in the light intensity of 150–200 μvolt, is about 1% of the transmitted light. A small constant D.C magnetic field is applied, in order to provide a quantum axis along the Rb light for the field independent F, $m_F = 0$–F', $m_{F'} = 0$

atomic transitions. It also provides some leverage for manipulating the hyperfine energy levels, so that the hyperfine transition frequency is exactly matched with the applied resonant microwave frequency. Besides, the D.C magnetic field takes care of any uncertainty in the buffer gas pressure. The constant D.C magnetic field is produced by a solenoid. The solenoid winding is done on a Hylem cylinder, a non-magnetic material. Its diameter is 70 mm and its length is 120 mm. The winding is compact, so that the good homogeneity of the magnetic field is maintained. A field of 250–450 milligauss is produced by the solenoid. A precise constant current source controls the current in the solenoid. In Figs. 1.5 and 1.6, the block diagram of

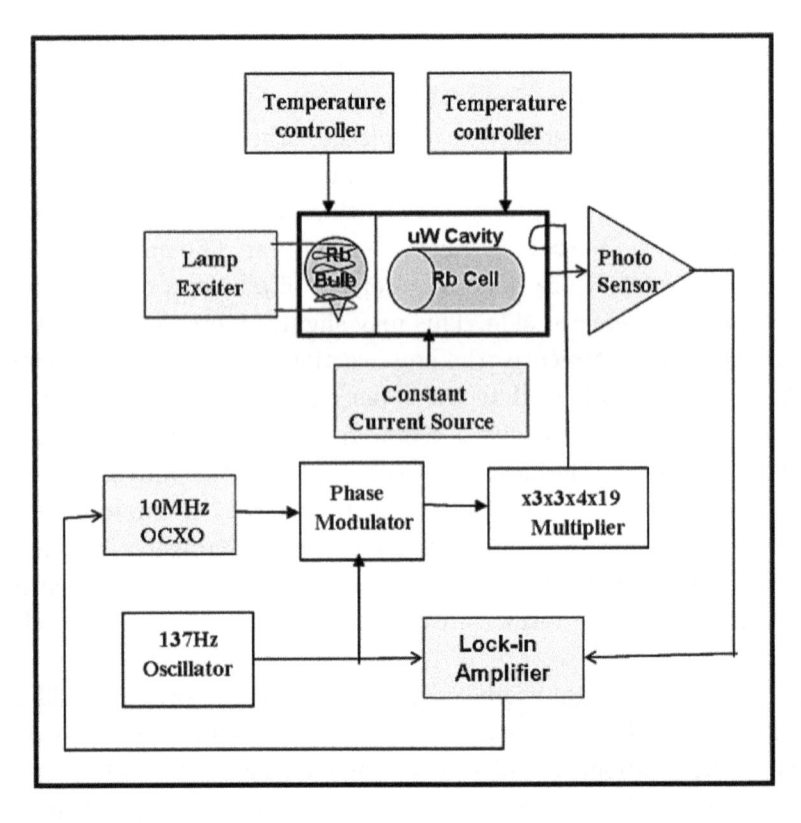

Fig. 1.5 Block diagram of Rb atomic clock (IFT).

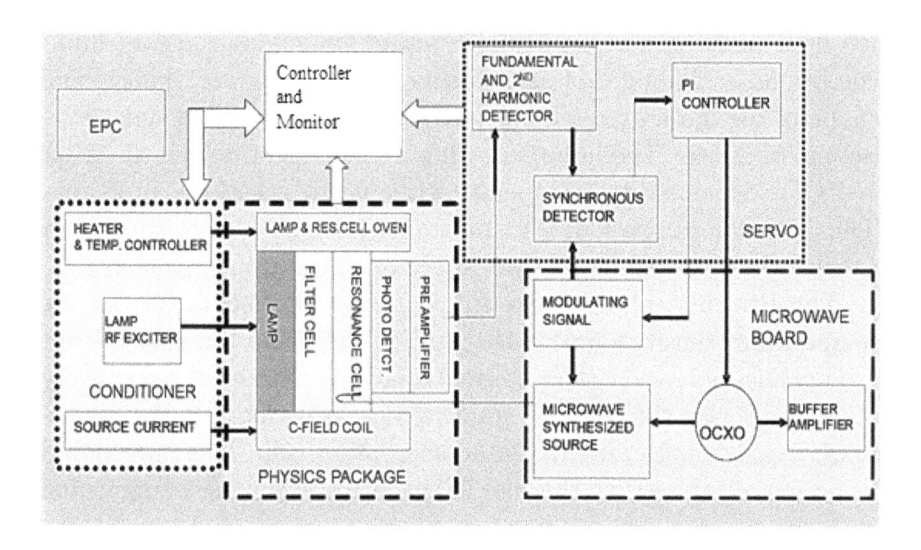

Overview of Rb Atomic Clocks and the Space Specifications 27

Fig. 1.6 Block diagram of Rb clock (SFT).

Rb atomic clock is given, for IFT and SFT respectively. The various components of the Physics package are described below.

1.4. Rb lamp

The Rb electrode-less lamp is the single most important part of the Rb atomic clock. The quality of the lamp decides the life of the Rb atomic clock. It is observed that many navigation satellites developed snag or could not last their expected life span, just because the Rb lamps failed. Therefore, thorough characterization of Rb lamp, using the calorimeter during its processing is very important. The glass manifold and UHV system used in filling these lamps should be ultra-clean. The type of glass used in bulbs should be carefully selected for low He permeability. The bulbs are typically spherical (5–10 mm diameter). The glass-blown bulbs are filled with a few hundred microgram of Rb and a starter/buffer gas, at a few milli-bar for easier ignition of the plasma. For excitation, Rb discharge lamp [21] is inductively coupled to an external coil of the exciter circuit. These lamps operate very well for compact clocks, due to the intrinsically correct light frequency for the optical pumping of the Rb atoms. The

Rb light bulbs are experimentally tested for a wide range of input conditions, including excitation frequency, input power, quality and shape of the inductive coil and temperature, for a stable and a low-power operation. The lamp, emitting a stable light output at about 100°C, is demonstrated to operate without any functional problems. Thus, the lamps are found to be stable and robust; ensuring their worthiness for a highly compact space qualified atomic clocks.

The Rb bulb, filter/absorption cell, and photo-detector, all are temperature sensitive and each requires a specific temperature for the optimal operation. The thermal analysis of Rb clock is critically done, to provide the required stable temperature for both the assemblies, i.e., Rb bulb and Rb filter/absorption cell. This is achieved using mainly the active thermal PID control techniques, augmented with the electrical passive heaters. The selected thermal control system ensures, that the temperature of each assembly is maintained stringently.

1.5. Frequency synthesizer

A low phase noise high resolution frequency synthesizer is needed to implement atomic interrogation. In this section, we describe the construction and evaluation of a low phase noise synthesizer for the 6.834 GHz resonant microwave frequency. The phase noise measurements at 6.834 GHz, confirm that there is no significant degradation of the 10 MHz local reference oscillator. The high stability quartz guarantees the frequency stability of the clock better than 5×10^{-12} $\tau^{-1/2}$. The performance of 10 MHz OCXO is characterized by referencing it to a Caesium atomic clock. The analysis, including the actual behaviour of different types of quartz-crystal oscillators, referenced to Cs atomic clock, is critically done. The best among all the OCXOs is selected as a 10 MHz source for the clock. The resonant microwave frequency at 6.834 GHz is required to be generated with 10 MHz OCXO as a reference source. The total multiplication factor of 684 is realized in $\times 9 \times 4 \times 19$ sequence. The detailed experimental measurements of the performance of a 10 MHz $\times$ 9 frequency multiplier are based on idler networks. The effect of 9th harmonic input impedance on the conversion gain is simulated and experimentally verified. A systematic design of a 360 MHz × 19, SRD based comb generator, is given under the chapter on the frequency synthesizer. The detailed analysis of the input matching circuits, impulse generator circuit and output matching circuits, is done for obtaining the minimum power of −26 dBm at 6.840 GHz and the pulse width less than 1.0 ns. The parameter-based model is first created for the SRD, DH543-62A, and its feasibility is verified by a test circuit. The output power dependence on the input power at 19th harmonics and the spectral purity (noise floor), are characterized for SRD based comb generators. The comb generator is fully characterized by varying the input power over the range of 20 dBm ± 5 dB and measuring corresponding output power at 19th harmonics. The variation of noise floor close to the output frequency (± 1MHz) is also characterized for different values of the input power. Then the effect of inserting transmission line between SRD and the filter circuit is studied. The simple generic advance design system (ADS) model for SRD and all passive components of a 360 MHz to 6.84 GHz multiplier, provides, accurately, the actual circuit operation for the output frequency. The variation of transmission line length between SRD and output filter circuit, that influences the 18th, 19th and 20th harmonic contents, is analysed by the simulation and experimental method. The design of a 6.834 GHz two-stage MIC amplifier with a small-signal gain of 24 dB is discussed. The amplifier, based on GaAs HEMT (high electron mobility transistor) technology with $f_T = 12$ GHz and $f_{max} = 20$ GHz is designed. A cascade configuration in combination with MIC (microwave integrated circuit) technology is adopted in the design. The proper characterization and careful optimization of the synthesizer constituents result in a low flicker phase noise of −85 dBc/Hz at 1Hz offset frequency. This corresponds to the frequency stability better than 5×10^{-13} at 1s, that is one order of magnitude better than the level estimated for the clock.

A microwave cavity and its magnetic field configuration play very important role in the Rb clock. Traditionally, the microwave cavity for passive Rb frequency standard is a cylindrical resonator that operates in either TE_{011} mode or TE_{011} mode. The reflection and transmission characteristics, based on the feed loop inserted into cylindrical metallic cavity, are simulated and experimentally verified for these modes. The influence of loop size on resonant frequency of the cavity modes is taken into account. The obtained results, for a loop inside the cavity as a feed, show that the values of resonant frequency, mainly, depend on loop wire dimensions, which are related to the loop impedance. The HFFS simulation technique gives information about the dimensions and position of the coupling loop, for achieving the best source matching in the operating frequency range. In the present case, 8 mm loop dia is chosen to excite the microwave cavity with good results. In the design process using the EM solver, resonant frequencies of the other higher order modes, quality factors and EM field distributions of dominant modes of TE_{111} and TE_{011} are rigorously analysed. The versatile visualization of the field distributions gives much more insight into the behaviour of the EM fields associated with the each mode. The level of EM field, an important parameter in the microwave cavity design, is estimated and also experimentally verified. The influences of the dielectric materials i.e., absorption glass cell, Teflon ring, FR4 photodiode PCB, along with the coupling loop dimensions, on the resonance frequency and Q are analysed. This process is helpful in the design of the dielectric loaded cavity for Rb atomic clock. We find that the inclusion of these dielectric materials into the cavity distorts the EM field lines considerably, and degrades the Q and also shifts the resonance frequency. The theoretical analysis agrees very well with the simulated and experimental results. Based on the analysis of the dielectric-loaded TE_{111} cylindrical cavity, a miniature cavity Rb absorption cell assembly is successfully developed, with the volume of the microwave cavity being only 23 cm^3. Finally, it is shown that such a design makes a TE_{111} cavity very small while maintaining high magnetic field uniformity. It may be adopted for the space qualified Rb atomic clock.

The influence of the cavity pulling effect on the frequency stability of the passive Rb clock is analysed, and it is shown that the microwave power and cavity temperature affect frequency shift. However, It can be minimized under reasonable RF power and temperature stabilization conditions. Therefore, a TE_{111} type of microwave cavity with a low temperature coefficient (TC) is developed. The experimental temperature coefficient is of the order of $3\,\text{kHz}/^\circ\text{C}$. That results in a clock TC limit of around $4 \times 10^{-14}/^\circ\text{C}$. The realizable control of the cell temperature is at the milli $^\circ\text{C}$ level. That meets the TC of the clock. An analysis shows that the cavity pulling effect can be neglected under reasonable temperature stabilization condition. The minimization in the TC of the cavity is the result of the compensation of the positive TC of the dielectric ring in the cavity by the negative TC of the metal part of the cavity.

When the Rb atoms in the absorption cell are subjected to a resonant microwave RF field, the transitions between g.s hyperfine levels are induced, that tend to reduce the population difference built up by the optical pumping. In this process, the amount of the optical radiation transmitted through the absorption cell and that incident on the photo-detector is reduced. Phase or frequency (PM or FM) modulation of the resonant microwave frequency, with small modulation index at some low frequency $137\,\text{Hz}$, also results in the modulation of the intensity of the light reaching the photo-detector. The photo-detector output contains the modulating frequency and its second harmonic ($274\,\text{Hz}$). The amplitude of the modulated light intensity is proportional to the slope of the microwave resonance absorption line. By synchronously demodulating the amplified output of the photo detector, a DC error voltage, known as clock signal is obtained. From the absorption signal, an anti-symmetric error signal is derived and used in the servo-control system. This is an unambiguous error signal for a servo unit. There is a positive error signal if the resonant microwave frequency is lower than the atomic transition frequency and vice-verse. The phase sensitive detector or lock-in-amplifier processes the error signal, that is applied to Electronic Frequency Control (EFC) of VCXO. With a closed servo loop, the VCXO frequency is locked to the frequency of ^{87}Rb ground state field free hyperfine transition. The corrected VCXO frequency becomes the output of Rb atomic clock.

We describe a lock-in-amplifier (LIA) based on an analog phase-sensitive detection technique, for processing of atomic weak signals in low SNR environments. The full details are given in the chapter on lock-in-amplifier. A physical explanation is given for how the "lock-in" produces the desired feedback signal and shape the feedback discriminator signal, by using a simple open loop experiment. We, briefly, describe the analog lock-in detector and its main building blocks. The analyses of the design and implementation details of the major sub-circuits are also included. The SNR of the error signal from the Physics package is measured using lock-in instruments, like RF synthesizer and spectrum analyser.

We provide a simple physical explanation and describe the design and measurement of an integrated lock-in amplifier or phase sensitive detector (PSD), based on a suitable mathematical model. We have successfully developed a simple and highly sensitive analog lock-in amplifier, for the detection of very low-level optical signals, which has a dynamic range of 140 dB and is capable of recovering input signals in the nano-ampere range. The system is operated at 137 Hz and is capable of extracting signals as small as 100 nV from the ambient noise. The developed LIA includes a high gain photodiode array to convert the incident optical signals into electrical current signals. The photocurrent is then amplified and converted by a low-noise trans-impedance amplifier into a voltage signal. The signal is first, processed by a low-noise amplifier (LNA) and a band pass filter. Then the signal is mixed with a known reference signal (137 Hz) of the same frequency as the input signal. The amplitude and phase of the desired signal are extracted from the mixed signal. From the mixer, the signal, which is a DC component, is filtered out by a low-pass filter. The LIA consumes, on an average, a power of 1.3 W with a ±15 V power supply. The LIA enables the recovery of signals, 60 dB below the noise level (a voltage ratio of 10^3). In addition, the need of the low power consumption related to space grade clock applications, requires the use of low-power low-voltage electronic components. A precise power management is needed to extend the operation time. The analog lock-in amplifier is designed, according to the specifications of the space qualified components. The LIA functionality is verified, using a low frequency constant current source as the input, to quantify the amplitude and slope of the atomic absorption signal i.e., the discriminator signal.

The discriminator signal is used, to lock the OCXO and to stabilize its frequency. The stability of the lock is affected by the total gain (product of AC and DC amplifier gains) of the error signal and the time constant of the low-pass filter in the lock-in amplifier. Adjusting all of these variables to obtain a stable lock requires time. We find that the lock-in amplifier is able to produce an error signal, that can keep the OCXO stabilized for a reasonable length of time. We believe that under the right settings, the OCXO frequency can remain locked on the atomic transition for a much longer time.

We characterize the discriminator signal that demonstrates the line widths and SNR of the Physics package. We present experimental data, validating these two unique features of resonances and its relation to the short-term frequency stability. It summarizes the physical basis and root causes of the environmental sensitivity of the Rb frequency standard. An understanding of the physical mechanisms and the causes of the environmental sensitivity are of obvious concern for the development of RFS (Rb frequency standard), especially since it is intended for the space application. Therefore, the experimental characterization is essential, of all clock critical parameters and the principal factors, that contribute to RFS instability. The experiments show a number of design trade-offs that can affect the clock performance. After verifying the criticality of the design, we conclude that it can reach the goal of the stability $\leq 5 \times 10^{-12} \tau^{-1/2}$ and even 1×10^{-14} for a sample time $\tau \geq 10^4$ s. The simulation and experimental work, on the thermal and vibration tests of the Physics package along with all the electronics, is critically executed for the developing the space qualified Rb atomic clocks. The thermovacuum model is designed to show its space worthiness. The work reported here is for the space qualified Rb atomic clocks for satellite navigation systems. These devices have a volume about $406 \times 301 \times 160 \, \text{mm}^3$, the power dissipation of 35 watt. The short-term stability of $6 \times 10^{-12} \tau^{-1/2}$ for the integration times $1\text{s} \leq \tau \leq 1000\text{s}$, and a medium to long-term stability reaching the 2×10^{-13} level at 10^4s. The volume of the Rb clock can be reduced considerably for the satellite navigation systems. The basic work reported in the book is in the Indian context. However, the present status of the Rb atomic clocks used in other GNSS is also exhaustively included.

1.6. Performance specifications of rubidium atomic clock for satellite navigation systems

The indicative performance specifications of the Rb atomic clock are documented here. The actual specifications may be better than those mentioned as follows:

Initial frequency accuracy: $\geq \pm 1 \times 10^{-10}$.

Frequency drift: $\leq 5 \times 10^{-13}$/day

Frequency stability: The short-term frequency stability of RAFS output at any fixed temperature in the operating temperature range shall be as follows:

1s	5 × 10–12
10s	2 × 10–12
100s	5 × 10–13
1000s	5 × 10–13 or better

Phase noise: The maximum allowed Single Side-band spectral density induced on carrier not to exceed is indicated below: offset from F0 Level dBc/Hz

1 Hz	−85 dBc/Hz
10 Hz	−100 dBc/Hz
100 Hz	−125 dBc/Hz
1 KHz	−135 dBc/Hz
10 KHz	−145 dBc/Hz
100 KHz	−145 dBc/Hz

Frequency temperature stability: The frequency variation at any temperature slot of 20°C, in the range of −5°C to +30°C $\leq \pm 5 \times 10^{-13}$ °C.

Frequency magnetic sensitivity: The frequency variation due to magnetic field variations $\leq \pm 1 \times 10^{-12}$ /Gauss.

Harmonics: The level of harmonics must be less than -30 dBc within Frequency band of ± 200 MHz.

Spurious: The spurious signal within the band, around the reference frequency, ± 1 MHz ≤ -80 dBc. Outside the band the spurious and harmonics should be -60 dBc.

Power steady state: The Power consumption of the RAFS should be about 30 Watts under steady state of operation.

Turn on: The power consumption of the RAFS is not to exceed 60 Watts.

Power supply stability: $\leq \pm 3 \times 1^{-12}$.

Barometric sensitivity $< 2 \times 10^{-9}$/bar
Critical pressure $< 1 \times 10^{-5}$ mbar

Dimensions: < 200 mm $\times$ 150 mm $\times$ 125 mm.

Mass: RAFS mass <3.0 Kg.

Qualification test sequence (applicable after Design Verification Model): Initial electrical measurements at 25°C.
Temperature storage test 24 Hrs at each storage temperature.
Humidity storage test 95% relative humidity and +40°C for 24 hours.
Post storage test electrical measurements at 25°C.
EMI/EMC test as per MIL-STD-461/462 D.
ESD test 7 KV, Radiated discharge, 5 KV Point discharge and structural current.
Post EMC test electrical measurements at 25°C Dynamic tests Sine vibration 20–70 Hz, 20g Random vibration 20–100 Hz, 3dB/oct.
Mechanical shock, 100–600 Hz, 15 dB/oct.
Pre-thermo-vacuum electrical measurements 25°C.
Active thermo-vacuum cycling tests qualification temperatures. Vacuum level: 10^{-5} Torr. or better.

Cold and hot turn-on test to be in the acceptance range. Any Temperature slot of 20°C in the range −5°C to +30°C.

Qualification range ± 5°C margin over the acceptable level.

Temperature monitoring of high power devices must be carried out.

Short cycles of 2 hrs dwell followed by Hot soak for 24 hrs: electrical measurements to be taken at the end of dwell period. Cold soak for 24 hrs: electrical measurements to be taken at the end of dwell period.

Post-thermo-vacuum electrical measurements 25°C.

Final electrical measurements 25°C.

External visual inspection.

Implementation plan:

The functional block diagram of the rubidium atomic clock, is given in the Fig. 1.7, which consists of the following sections:

Physics package assembly:

Physics package assembly consists of following sub-sections:

- Rb Lamp
- Absorption cell in integrated and separate configuration
- Microwave cavity

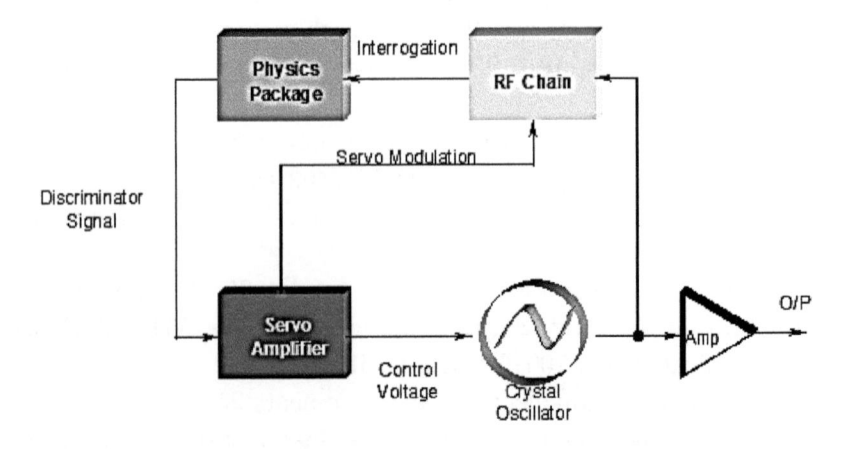

Fig. 1.7 Sectional schematic of Rubidium atomic clock.

- Magnetic field and temperature control
- Photo-detector for the atomic resonance detection
- Base Plate.

RF generation section:

RF section consists of following sub sections:

- Oven controlled VCXO
- 137 Hz modulator
- Freq. Synthesizer/Multiplier.

Servo section:

Servo section consists of following sub sections:

- Pre-amplifier
- Lock In Amplifier
- 137 Hz frequency generator.

EPC is also part of Rubidium clock, however its parameter and design are finalized after the competition of DVM phase.

The implementation plan is as per the description given below.

(A) *Design Verification Model (DVM)*

(i) *Physics package*

The Physics package of the Rb atomic clock consists of Rb Lamp, Rb absorption cell, microwave cavity, photo-detector, heater/C-field coils and magnetic shields. The interface of Physics package with other electronics circuits, microwave input for RF excitation, photo-detector output synchronous processing, C-field control current, lamp exciter coil input and heater inputs for the lamp and the absorption cell.

(ii) *Mechanical and thermal Design*

The mechanical and thermal design, simulation and analysis for the Physics package, are carried out to ensure the space worthiness of the Rb atomic clocks.

(iii) *Fabrication and Assembly*

Glass components are fabricated for the Physics package, mechanical parts for housing the cells, Al-alloy for microwave cavity, thermal isolation material for packaging of the Physics package components, heater coils, magnetic field coils, magnetic shielding.

(iv) *Testing*

The spectroscopic and micro-calorimetric tests related to the Physics package.

(B) *Engineering Thermal Model (ETM)*

After the completion of DVM model, ETM is developed for Rb atomic clock. R & Q A is done at the design phase of ETM model, for the component and raw material selection, for the space grade procurement, fabrication process qualification for the Rb absorption cells and other spin-off new technical area.

Summary

The timing precision provided by the atomic clocks, based on vapour-cell technology, plays an important role in the wide variety of scientific fields and human activities [12] such as measurement of fundamental physical constants, communications industry, military systems and navigation systems [12, 23–26]. The accurate positioning and navigation systems [27–30] (Global Positioning System (GPS), Galileo, GLONASS etc.), which consist of several satellites, transmitting synchronized coded signals, depend on the accurate atomic clocks in each satellite that ensure the synchronization of the signals. By measuring the time differences between the signals from the satellites, one can calculate its position and height to within a few meters or centimeters [31]. GPS is widely used in a number of scientific fields, including geology for measurements of Earth's crustal deformations and continental drifts, Palaeontology and Archaeology for recording the locations of fossils and artefacts, and civil engineering for monitoring the settling of man-made structures over the period of time and land surveying. GPS is having an impact on the lives of the common person as well, as more people are relying on GPS navigators while hiking and even while driving their cars. So the overall performance of navigation's payload depends on the performance of the onboard Rb atomic clocks [32].

Precise timing by the atomic clocks is also necessary for tele-communication industries, such as mobile, radio and television broadcasting and internet [33–35]. In addition to facilitating telecommunication, the atomic clocks help to manage the long-distance electrical power transmission [36, 37] across different regions of the country. Another field that has benefited from precise timing is that of Metrology. The accuracy and precision of the atomic clocks has made the Second, the most accurately realized unit of measurement among all physical quantities. Because of the reliance on the time unit "Second" for the definitions of several SI units, the ability to accurately measure time is important in many scientific fields [38, 39]. It also provides many opportunities in other scientific applications such as measurements of the variation in fine structure constant [32], test of relativity [40], and a precise spectroscopy. Besides scientific research, these clocks are crucial in other scientific endeavour such as Gravitational red shift experiments (known as the Gravity Probe-A (GP-A) experiment), Doppler wind experiment (DWE) (e.g., Cassini Huygens) [28, 41] and the precise measurement of the speed of neutrinos [42].

The Rb atomic clocks are also used in numerous military applications such as secure communications, electronic warfare, command and control, telemetry and navigation [27–30]. The modern military systems need clocks with good precision and high stability [23–24] over a wide range of physical parameters such as temperature, vibration, acceleration and radiation etc. The laboratory clocks like primary Caesium (Cs) fountains and optical clocks exhibit excellent stabilities of the order of $\sigma y(t) \leq 1 \times 10^{-13} \tau^{-1/2}$, but are bulky and expensive. Even cold atom clocks or the optical clocks have been proposed for the space applications with the target of $1\,\text{m}^3$ volume, $230\,\text{kg}$ mass, and $450\,\text{W}$ power consumption [26], hence a trade-off must be exercised between the stability and the portability. The requirement of compact size, low power consumption, very fast warm-up stabilization and cost effectiveness, is satisfied by Rb vapour cell clock, but not met by other atomic clocks. Recently developed portable standards, such as the passive Rb standard exhibit a reasonable trade-off with volume ($13\,\text{cm}^3 < V < 28\,\text{cm}^3$), mass ($5\,\text{kg} < m < 18\,\text{kg}$), power consumption ($25\,\text{W} < P < 80\,\text{W}$) and stability $7 \times 10^{-13} < \sigma_y\,(1\,\text{s}) < 1.5 \times 10^{-12}$ and $1 \times 10^{-14} < \sigma_y\,(10^4\,\text{s}) < 3 \times 10^{-14}$. The satellite based navigation system requires on-board ultra-stable precision Time and Frequency Standards to generate various frequencies and timing information. The Rubidium atomic clocks are universally used in the global navigation satellite systems for meeting the long-term requirement of the stable and accurate Time.

Chapter 2

Theoretical Aspects of Rb Atomic Clock

> In this chapter the theoretical discussions on the atomic structure and energy levels of the Rb atoms are included for providing insight into their suitability for the atomic clock. The population inversion through the Optical Pumping is also discussed, as it is the main process used for creating the required population inversion, and in turn generating the clock signal.

2.1. Atomic structure of Rb atom

The energy levels of Rb, an alkali metal can be described using Paschen notations. In Rb atomic clock, the g.s hyperfine field free sub-levels are selected for getting the atomic clock error signal, which is applied to EFC of a VCXO. Figure 2.1 shows the hyperfine structure of ^{87}Rb for the ground and P excited states. In particular, we describe the transitions between the sub-levels of the 5S–5P states. To understand the energy levels of Rb, we need to consider the coupling between the electron and the nucleus in low magnetic field. The angular momentum of the valence electron is given by $J = |L \pm S|$, where L is the orbital angular momentum and S is the spin angular momentum. For the hyperfine energy levels structure, the total angular momentum F of the atom also includes the nuclear spin, which is denoted as I. The atomic angular momentum is then given by $F = |J \pm I|$. The standard spectroscopic notations describe the electronic state as $n^{2S+1}L_J$, where for g.s of Rb, the n = 5 is the principal quantum number of the outermost electrons. In the superscription 2S + 1, the values of L are denoted by letters: 'S' for L = 0 and 'P' for L = 1. For the ground state of ^{87}Rb, the orbital angular momentum of the outer electron is zero (L = 0), thus only its

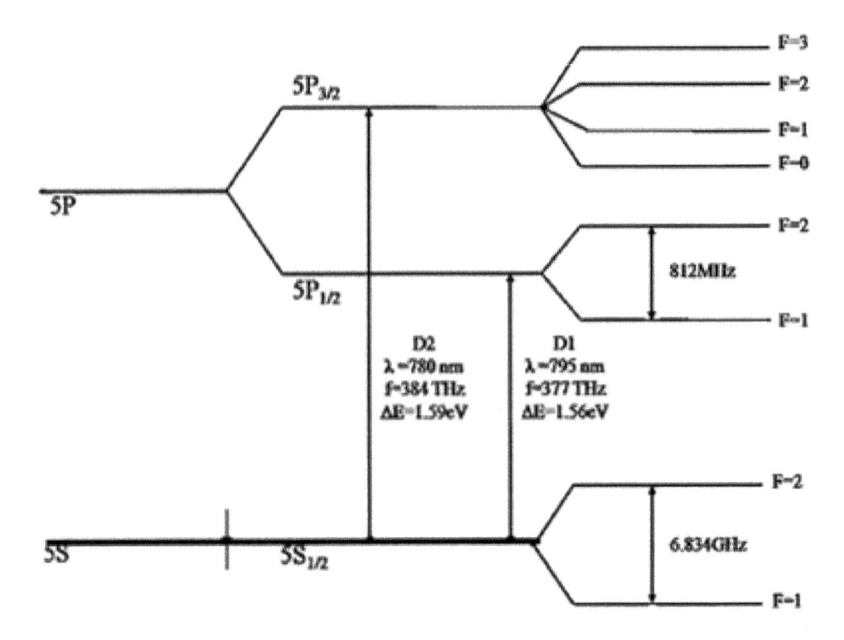

Fig. 2.1 The energy level diagram of the 5th electron shell of ^{87}Rb for the ground and first excited states.

spin momentum S = 1/2 contributes to make the total electron angular momentum, J = 1/2. Due to this spin-orbit coupling, the first excited state P has L = 1 and S = 1/2, so we have J = 1/2 or 3/2. The excited state P splits up into $5^2P_{1/2}$ and a $5^2P_{3/2}$ states. This is according to the rule of angular momentum addition, J = L + S and $|L - S| \leq J \leq |L + S|$. The radiation corresponding to the energy difference between the $5^2S_{1/2} - 5^2P_{1/2}$ levels of Rb is termed as the D1 line; its wavelength is roughly 794.8 nm. The $5^2P_{3/2}$ state is separated from the ground state by an energy gap equivalent to 780 nm wavelength; it is called the D2 line. The hyperfine structure arises due to the coupling of the electron angular momentum J with the total nuclear angular momentum I. This leads to the splitting of the ground state and the excited states. The total angular momentum F is given by $F = |J \pm I|$, where the range of the magnitude of F is $|J - I| \leq F \leq |J + I|$. The angular momentum I of the ^{87}Rb nucleus is 3/2, and so the allowed values of F for the $5^2S_{1/2}$ and $5^2P_{1/2}$ states of this isotope are $3/2 \pm 1/2$. Similarly, for the excited state, $5^2P_{3/2}$ J = 3/2 and I = 3/2 and F can take any of the values 0, 1, 2 or 3. Therefore, for the ground state hyperfine splitting, there are two levels, corresponding to F = 1 and F = 2. These levels are separated by 2.826598×10^{-5} eV, or 6.834683 GHz as shown in Fig. 2.1. The Zeeman level splitting occurs in the presence of a weak external DC magnetic field B. When an external magnetic field is applied, each of the hyperfine energy levels splits into $2F+1$ magnetic sublevels (Zeeman splitting). Figure 2.2, shows the magnetic sub-levels and the allowed transitions in the ground state of ^{87}Rb. The application of a magnetic field, lifts the degeneracy between different Zeeman sublevels of the states with the same total angular momentum F, but different projection m_F along or perpendicular to the quantization axis, defined by the magnetic field.

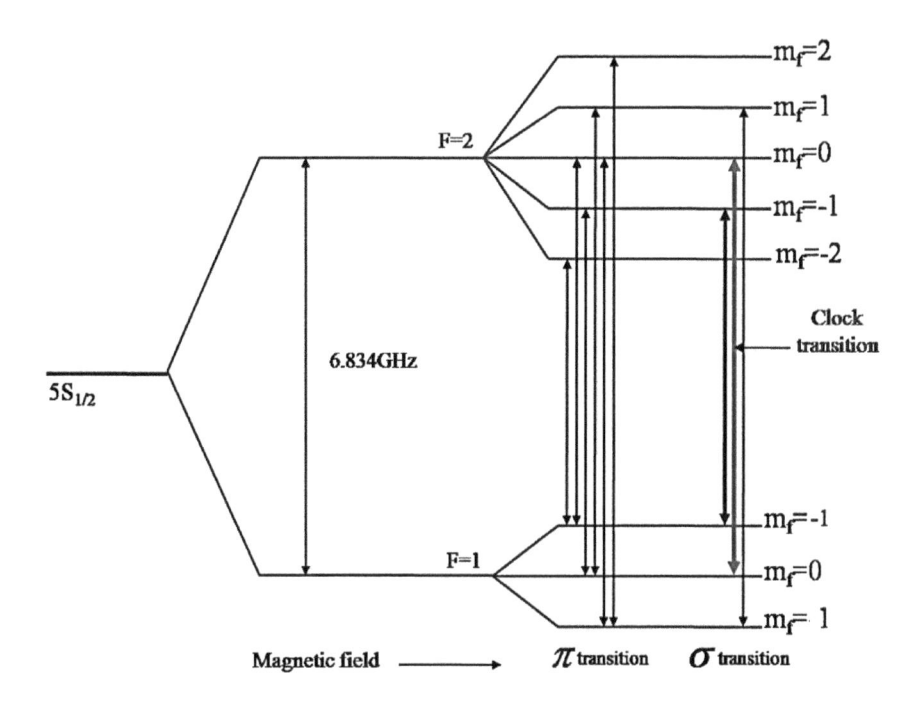

Fig. 2.2 The magnetic sublevels and allowed transitions in the ground state hyperfine of ^{87}Rb.

In magnetic dipole transitions, if the oscillating magnetic field is parallel to the static magnetic field, which is also the atomic quantization axis, $\Delta m_f = 0$ transition (π-transition) is allowed, and if the oscillating magnetic field is perpendicular to the static magnetic field, $\Delta m_f = \pm 1$ transitions (σ-transition) are allowed. The hyperfine levels F = 1 and F = 2 consist of three and five magnetic sub-levels respectively. Among the three π transitions, only F = 1, $m_f = 0$ < - > F = 2, $m_f = 0$ is used in the Rb atomic clock. This has second order dependence on the magnetic field and is known as field independent transition.

The Zeeman splitting for σ-transitions is proportional to the magnetic field, for small field strengths, typically less than a Gauss. This splitting gives rise to Larrmor precession with frequency, $\omega_L = \Delta E_L = \gamma B$. Here the gyro-magnetic ratio γ, for the alkali atoms is given by $\gamma \approx \pm 2\pi \times (2.8\,\text{MHz/G})/(2I+1)$, where I is the nuclear spin and the sign corresponds to the hyperfine manifold. The F = I+1/2 manifold yields a positive sign in the gyromagnetic ratio. However, for large magnetic fields, Zeeman energy level splitting is non-linear and is given by the Breit–Rabi splitting formula. We can calculate the splitting by studying the ground state Hamiltonian of the alkali atom, which is expressed as:

$$H = A_J I \cdot J + g_s \mu_B S \cdot B - g_I \mu_N I \cdot B, \qquad (2.1)$$

where μ_N is the nuclear magneton, $A_J = 2\omega_{hf}/(2I+1)$ is the hyperfine coupling constant describing the strength of the hyperfine interaction, $g_s = 2$ is the electron g-factor, $\mu_B = 9.274 \times 10^{-24}$ J/T is the Bohr magneton, and g_I the nuclear g-factor. The energy spectrum can be calculated from the eigenvalues of the Hamiltonian (A. Corney, Atomic and Laser Spectroscopy, Oxford, Clarendon, 1977). The energy levels are expressed as:

$$E(F = I \pm 1/2, m_F) = -\frac{\omega_{hf}}{2(2I+1)} - g_I \mu_N B m_F$$
$$\pm \frac{\omega_{hf}}{2}\sqrt{x^2 + \frac{4x m_F}{2I+1} + 1}, \qquad (2.2)$$

where

$$x = \frac{2(g_s\mu_B \div g_t\mu_N)B}{(2I+1)A_J} = \frac{(g_s\mu_B \div g_t\mu_N)B}{\omega_{hf}}. \quad (2.3)$$

Notice that the energy levels spacing, as a function of the magnetic field, is now non-linear. In the low field, the adjacent sublevels are separated by $g_I\mu_N B m_F$. This clearly scales linearly with the magnetic field. In the high field regime, the last term in Eq. (2.2) causes the splitting which is not directly proportional to the magnetic field.

2.2. Optical pumping

The Nobel laureate Alfred Kastler devised the optical pumping method [18, 19], a pioneering work, which resulted in the development of the Rb atomic clock. As the name suggests, in this technique the optical radiation is used for exciting the atoms. The optical pumping opened up a new field for studying hyperfine structures of Alkali atoms. The optical pumping method is used for exciting atoms, particularly Alkali, to higher energy levels from the ground hyperfine levels. These levels are more or less equally populated, as per the Boltzmann distribution, at a given temperature, due to small energy gap among them. These optically pumped excited atoms, undergo spontaneous emission and as per the selection rules, populate the ground state hyperfine sub-levels of interest. In this process, the atoms have population inversion in the ground state. This population inversion prepares the atoms for interrogation and getting the necessary information, thus resulting in various exciting applications. The Rb atomic clock is one of the ardent examples of the very effective and useful application of the optical pumping technique. Alfred Kastler in 1952 experimentally demonstrated RF resonance in Alkali atoms using optical pumping technique. Thus, establishing and also successfully validating the theory of his very powerful optical pumping technique. His pioneer work of optical pumping also resulted in development of lasers, atomic clocks, magnetometers and many other experimental tools.

In the case of ^{87}Rb atom, the ground state hyperfine levels F = 1 and F = 2 have very small energy gap and as a result both levels have more or less equal population under thermal equilibrium, as per Boltzmann distribution. The optical pumping technique, described above, creates necessary population inversion between the levels of interest. The intensity of the light transmitted through the sample of ^{87}Rb atoms with some inert buffer gas at a few Torr pressure in an absorption cell, after filtering by ^{85}Rb atoms, initially shows marked reduction due to the absorption by ^{87}Rb atoms in F = 1 level. The ^{87}Rb atoms have optically induced population inversion and F = 2 hyperfine level becomes fully populated at the expense of F = 1 level. The transmitted light absorption reduces and the ^{87}Rb vapour becomes transparent. The light intensity passing through the Rb vapour regains its maximum value. However practically, the level F = 2 may not become fully populated. The final state of population inversion depends on factors such as the intensity and spectral profile of the pumping light, the relative transition probabilities of all allowed states, relaxation rates through inert buffer gas or combination of gases, spin exchange collisions and wall collisions.

From the above discussion, it is quite obvious that optical pumping is a very effective technique, using the optical radiation, for creating population inversion between two desired energy levels of any ensemble of atoms. The population inversion is required to interrogate the atoms for developing state-of-the-art devices. The goal of optical pumping [43] is to preferentially alter the thermal distribution of populations among the hyperfine/Zeeman levels in the ground state of atoms. In the present case, a ^{87}Rb discharge lamp emits optical radiation that contain D1 (795nm) & D2 (780nm) spectral lines of Rb. Therefore, by a natural coincidence, the ^{85}Rb isotope absorbs, Fig. 2.1b, the spectral line arising due to the transitions from $F = 2$ to P-levels of ^{87}Rb. In this process, the filtered radiation incident on ^{87}Rb resonance cell preferentially excites atoms out of the F = 1 states [44]. The atoms then decay spontaneously, at approximately equal rate $\Gamma/2$, to the two hyperfine ground states. After several absorption-emission cycles, the population inversion is created and the atoms accumulate in F = 2 state at the expense of F = 1 level, which gets depleted. In this condition, the tuned 6.834 GHz microwave cavity is excited exactly at the hyperfine transition frequency. The microwave interaction repopulates level F = 1 level. As a result, the light is again absorbed by ^{87}Rb atoms, and the intensity of the transmitted light shows a dip at the photodiode detector. This dip in the intensity is used for generating the clock error signal.

2.3. Microwave-optical double resonance

The principle of operation of the Rb atomic clock is based on using an atomic transition frequency as a reference to stabilize the frequency of a quartz oscillator (VCXO). In double-resonance (DR) in the Rb atomic clock [45], the optical pumping light from a ^{87}Rb discharge lamp, which contains D1 and D2 lines, illuminates the integrated resonance cell or an absorption cell in separate filter cell technique. In the integrated cell, ^{85}Rb isotope component in natural Rb filters out the ($5^2S_{1/2}$ state, $|F = 2, m_f = 0\rangle$ to $5^2P_{1/2,3/2}$ states) spectral line from the ^{87}Rb lamp to create the necessary population inversion in the ^{87}Rb hyperfine states in the resonance cell. This is because the F = 3 state of ^{85}Rb happen to be matching naturally with the F = 2 state of ^{87}Rb. Hence enabling natural filtering of undesired radiation line. This optical pumping alters the thermal distribution of populations among the Zeeman states and creates a ground state polarization in ^{87}Rb atoms. The clock transition ($5^2S_{1/2}$ state, $|F = 2, m_F = 0\rangle$ to $|F = 1, m_F = 0\rangle$) is then detected by applying a resonant microwave field to the atoms via a microwave-cavity, which is kept in the absorption cell. The absorption of the light as a function of the microwave frequency is a measure of the atomic ground state polarization. The frequency of a voltage controlled quartz oscillator is locked to this absorption signal, using a phase-sensitive detection. The continuous phase-correction voltage from the phase sensitive detector helps in actualizing a highly precise time and frequency signal from the phase locked quartz oscillator.

2.4. The space qualified Rb clocks

The satellite navigation is an important resource for all countries but is controlled by a few countries. There are reasons that these navigation facilities may not be available to all under adverse circumstances. Therefore, in order to fill the gap between GNSS systems of today and future possibility of restricted satellite navigation, many countries are establishing their own satellite navigation systems to provide location and time information to regional or global users as an independent navigation systems other than GPS , GALILEO and GLONASS, with a constellation of GEO and GSO orbits. In 2008, Space applications centre in Ahmedabad, India proposed to Indian Space Research Organisation (ISRO) a joint project in collaboration with National Physical Laboratory New Delhi India (NPLI) for the first experimental study on the development of Rb clocks for IRNSS (Indian Regional Navigation Satellite System). The project included activities on the design verification model (DVM) as well as on the development of qualified model (QM) of the Rb atomic clock. The study and design of various sub-systems of DVM & QM models of space qualified Rb clock and in particular, realization of the compact, reliable and low power space qualified Rb clock was undertaken. The Rb atomic clock is being space qualified by rigorous testing and integration of the sub-systems. The R&D work provides important technological and scientific inputs for developing the Rb clock for applications on the ground and in Space.

Chapter 3

Studies on Rb Lamp and Driver Oscillator

> The electrode-less Rb lamp is the most critical component of the space Rb atomic clock. Many satellite navigation missions failed because of the Rb lamps going bad or stopped functioning. Therefore, in the book we include all aspects of Rb lamp and its exciter circuitry in a very detailed and elaborate manner.

3.1. Critical role of Rb lamp

Rubidium atomic clocks typically use Rb electrode-less discharge lamps to supply the optical radiation for the optical pumping, a process central to the operation of Rb clocks. The Rb lamp is also the most critical part of Rb atomic clock's Physics package. An exhaustive study on Rb lamp is reported in this chapter. These studies on the Rb lamp are very important for the industry manufacturing space qualified Rb atomic clocks. For the overall life of the satellite navigation system, the atomic clock's reliability is of the utmost importance. Consequently, Rb discharge lamp reliability and its life are of great concern. One of the goals of the space Rb atomic clock's research and development program is to ensure the reliability and long life of the lamp in the satellite compatible Rb atomic clock.

The vapour Rb discharge lamps are used as the optical sources, having a defined spectral content, for optical pumping and absorption in Rb atomic frequency standards. The Rb plasma is essentially a low pressure Rb-Xenon discharge, taking place in a spherical glass bulb. The discharge plasma, composed of nearly equal number of Rb ions and free electrons, is powered by the RF field [46]. The lamp requires a voltage gradient in the gaseous plasma in order to impart energy to the free electrons of the plasma. This gradient is produced by a changing magnetic field in the inductive coil [47, 48]. The Rb vapour density is very low at room temperature ($\sim 10^{-6}$ millibars of vapour pressure at 30°C), hence requiring an extremely high field strength to ionize the Rb atoms (Paschen curve) [49]. The temperature of the lamp is also increased up to 110°C by using a resistive heater, for increasing the Rb vapour density in the cell. To reduce the RF field and hence the electrical power to ignite the glow discharge, an inert gas at low pressure is added to the bulb. The bulbs are filled with the noble gas Xenon or Krypton, at 2.0 ± 0.2 torr, with 99.995% purity. Xenon, used as starter gas, has a relatively high mean free path among noble gases, with an ionization potential of 12.12 eV. Such vapour discharge lamps are ignited by an RF excitation field, generated by a coil that is driven by a RF oscillator or exciter. This type of RF power oscillator can reliably start the lamp and maintain constant lamp output under varying conditions. The fluctuations in the lamp output due to the temperature change and variations in the excitation power are the difficulties often encountered in starting and sustaining an electrode-less vapour discharge lamp, using conventional excitation circuitry. In addition, some variations in the oscillator power supply, such as low frequency ripple, can vary the light output. However, the most critical problem is the variation in the load presented to the oscillator, by the vapour discharge in the Rb lamp, that can induce variations in the excitation power and thereby cause periodic fluctuations in the lamp output. The electrical impedance of RF discharge lamp depends on the gas pressure, its composition, on the geometry and material of the coil, on the frequency of excitation and the temperature of the lamp. The impedance of the Rb bulb also traverses extremes, from a pure capacitance prior to the ignition and to a complex impedance after ignition. Therefore, it is obviously difficult to estimate the bulb impedance theoretically. The 2–3 watt power RF oscillator is used to excite the Rb bulb. All the parameters of the oscillator circuit such as the absorbed RF power by the bulb, operating oscillation frequency, loading due to sill fill epoxy material and operating mode of the bulb, are experimentally characterized.

The objectives of this chapter are important on two aspects. Firstly, based on the experimental characterization with a power amplifier and the theory of LC resonator, a simplified ADS compatible behavioural model of the Rb bulb load is developed. This emulates the electrical response of the inductive coupled electrodeless Rb lamp oscillator circuit. Secondly, the effect of the coil load in the oscillator circuit is also investigated. This methodology relies both on experimental observations as well as on the theoretical reasoning. The modelling of Rb bulb exhibits two major electrical features of the electrode-less lamp operating at RF: the dependence of the lamp's impedance on the power level and its dynamic response to changes in the electrical excitation. Once developed, the model is calibrated against the experimental data and also verified by the independent measurements. This proposed circuit model is useful in the exciter circuit's frequency domain analysis and its simulations as well. The frequency domain analysis and design consideration are discussed in Sec. 3.2. Finally, the simulated and experimental results are provided in Sec. 3.3 to verify the model and frequency domain analysis results.

3.2. Characterization of RF discharge Rb lamp

For the electrode-less Rb lamp, the power is delivered to the lamp either by the inductive or capacitor coupling. Presently, many of the space qualified electrode-less Rb lamps use inductive coupling and the low operating frequencies, as the power conversion is the most efficient. The Rb bulb and inductive coil form a parallel resonant circuit as depicted in Fig. 3.1. The precise impedance of the bulb needs to be determined for developing the electrical lamp oscillator circuit, for igniting RF discharge. For measuring the load of the Rb bulb, a series LC resonant circuit is chosen for driving the Rb discharge lamp. The LC stage refers to the load stage, that constitutes the series LC resonant circuit and the Rb bulb.

The performance of the lamp exciter circuit can be efficiently optimized at the design stage, through accurate simulation with the realistic equivalent circuit model of each component.

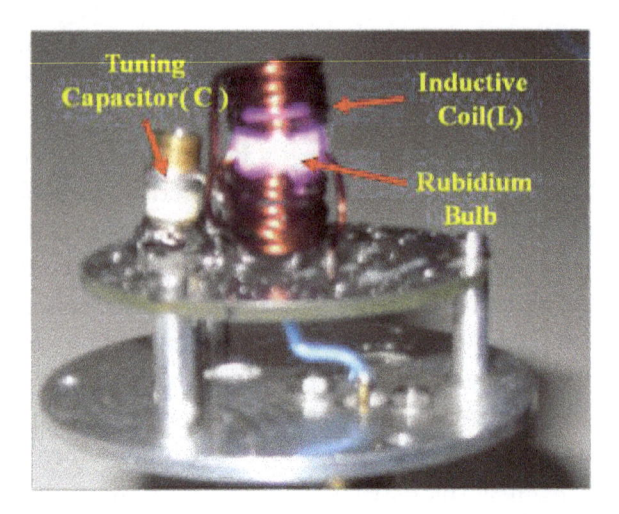

Fig. 3.1 LC resonator circuit with Rb bulb.

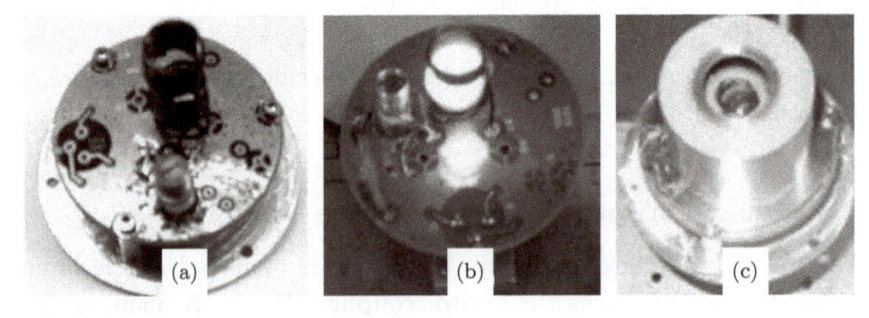

Fig. 3.2 LC resonator circuit (a) Only L&C (b) L&C with Rb bulb (c) L, C and Sill fill material.

3.2.1. *Studies on inductive coil with Rb bulb*

It is necessary to evaluate the coil parameters, such as self-inductance, impedance and self-resonance frequency for driver circuit. The values, self-inductances, internal resistances and self-resonance frequency (SRF) of the coil are confirmed by a one-port measurement method. The single port S-parameter is measured to be in a frequency range from 30 MHz to 150 MHz using a Vector Network Analyser (VNA). The entire assembly of LC circuit with the Rb bulb is shown in Fig. 3.2.

Studies on Rb Lamp and Driver Oscillator 53

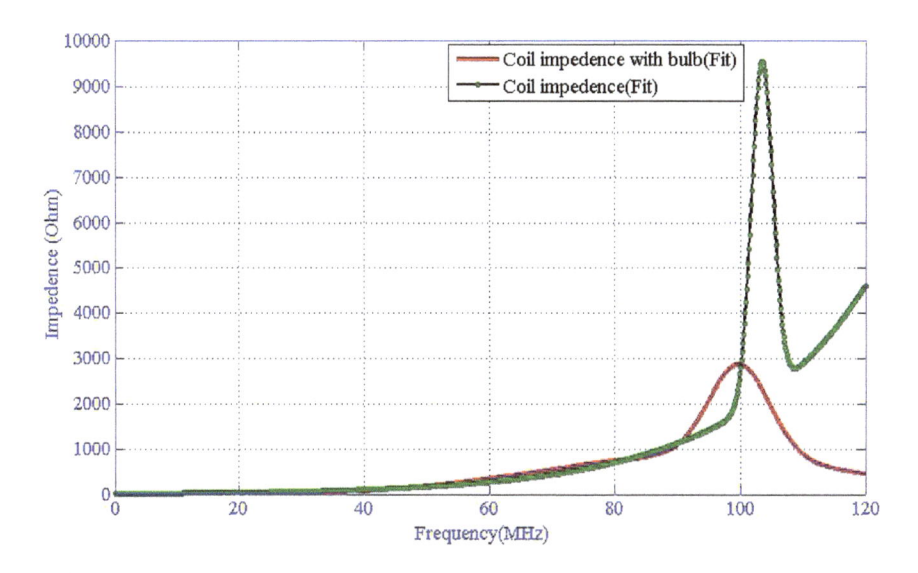

Fig. 3.3 Coil impedance vs frequency plot with and without bulb.

The one port S-parameter measurement for the coil module is used in the modelling the Rb bulb, and for maximizing the performance of the lamp driver or exciter circuit. The inductance and the impedance curves are plotted and shown in Fig. 3.3.

As a result of the Rb bulb loading, the coil inductance and impedance are reduced in both the cases. The measured inductance is 11 uH at 105 MHz. When we put the bulb into the coil, the self-resonance frequency of the coil is shifted to 100 MHz with the inductance reduced to 4.5 uH. The impedance is also reduced from 7.2 kΩ to 2.8 kΩ at 100 MHz. The resonance frequency is reduced on putting the bulb into the coil. This is a theoretical verification that on the dielectric loading of the bulb, its capacitance is added parallel to the parasitic capacitance of the coil inductor.

3.2.2. Studies on LC resonant circuit loaded with Rb bulb

Figure 3.4 refers to the load stage that constitutes the series LC resonant circuit. The impedance spectrum of the LC parallel resonant circuit, with Rb bulb and without Rb bulb is accurately measured using

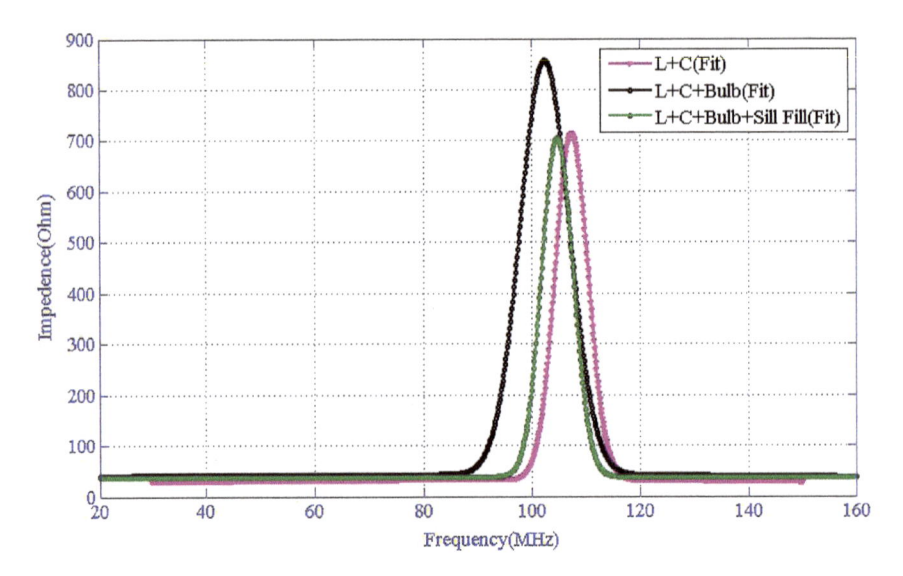

Fig. 3.4 Frequency vs LC impedance plot without bulb (purple) with bulb (black) plus sill fill (green).

a PNA-X Agilent N5244A network analyzer (10 MHz–42.5 GHz). The Rb bulb is placed inside the coil and the aluminium block with thermal sill fill epoxy material, as shown by green colour in Fig. 3.4. The Rb bulb with sill fill material adds a capacitive load, whose precise impedance needs to be determined for developing the electrical drive circuit. The impedance magnitude at resonance of an RLC circuit along with Rb bulb and sill fill material is also shown in Fig. 3.4.

The values of LC can be optimized for the maximum voltage transfer. The impedance of these components is measured using the vector network analyzer (VNA). For a 11μH inductor and 105 MHz drive frequency, the appropriate bulb capacitance value is calculated to be 0.6–0.4pF. Higher inductance values also entail higher skin and proximity effects, leading to higher resistive losses. In order to bring the required inductance value to an acceptable range, a tunable capacitor, with a higher capacitance than the bulb load, is added in parallel with the bulb load, to reduce the overall load reactance. The bulb load is found to be almost fully capacitive with a capacitance value of 0.4–2.2 pF and a total impedance of about 700 Ω at 100 MHz.

3.2.3. S-parameter and harmonic balance simulation

The S-parameter is an efficient means of analysing the linear networks using RF or microwave frequencies. In the S-parameter simulation [50] matched loads are used and no open or short circuit condition are applied to analyse the linear circuits. We use the S- or scattering parameter simulation to study the lamp RF exciter circuit. The voltage ratios are in dB at input and output ports. This terminations in the RF and microwave regions are easier to use thus making the detailed studies on the network easier and faster. In the present case, the RF coil in the lamp circuit is analysed using single port S-parameter. A two port S-parameter may be reduced to single port and one needs to evaluate the reflection coefficient S11, which is the ratio of the reflected power to the total input power. In a two port S-parameter ADS simulation, we have a 2 × 2 matrix, Eq. (3.1) with components S11 and S22 giving the forward and reverse reflections input and output matched impedances respectively. S12 represent the forward transmission loss or gain, and S21 describes reverse transmission leakage or isolation. The two port S-parameter is very versatile, as on it, higher ports analysis can be built. The S-parameter matrix components are measured using vector network analyser (VNA).

$$\begin{bmatrix} S11 & S12 \\ S21 & S22 \end{bmatrix} \quad (3.1)$$

For the frequencies in the microwave region and the non-linear circuits, the harmonic balance simulation is very useful. Therefore, the microwave resonant frequency or synthesizer signals discussed in Chapter 5, are simulated and analysed using the harmonic balance.

3.2.4. Effect of sill fill epoxy on LC resonant circuit

The lamp assembly with sill fill material and thermal heater is shown in Fig. 3.5. The entire assembly has the bulb with sill fill material, placed in an aluminium cylinder. Using the VNA, the loading effect into the LC resonator due to the Rb bulb and sill fill material is measured.

Fig. 3.5 Bulb assembly with thermal epoxy and heater.

By observing the values of S-parameter, it is noticeable that the value of input return loss (S_{11}) decreases when the bulb is added with sill fill epoxy material. This agrees with the theory that the S_{11} decreases as more 'impedance mismatch' occurs when the bulb and coils are fixed with the sill fill epoxy. The S_{11} in the case, (c) resonator with bulb and sill fill epoxy, is about 1.5 dB, which is smaller than that in the case (b) LC resonator with bulb, $S_{11} = 7$ dB and case (a) only LC resonator, $S_{11} = 2$ dB , at 78 MHz, in Fig. 3.6.

By analysing the experimental data with the network analyser, the bulb impedance and its capacitive load are calculated. A model of the Rb bulb is developed with the ADS simulation software.

3.2.5. *Modelling of Rb bulb in ADS software*

The circuit simulation using Spice or Saber or Simulink as the simulator is reported here, as these are the indispensable tool in the

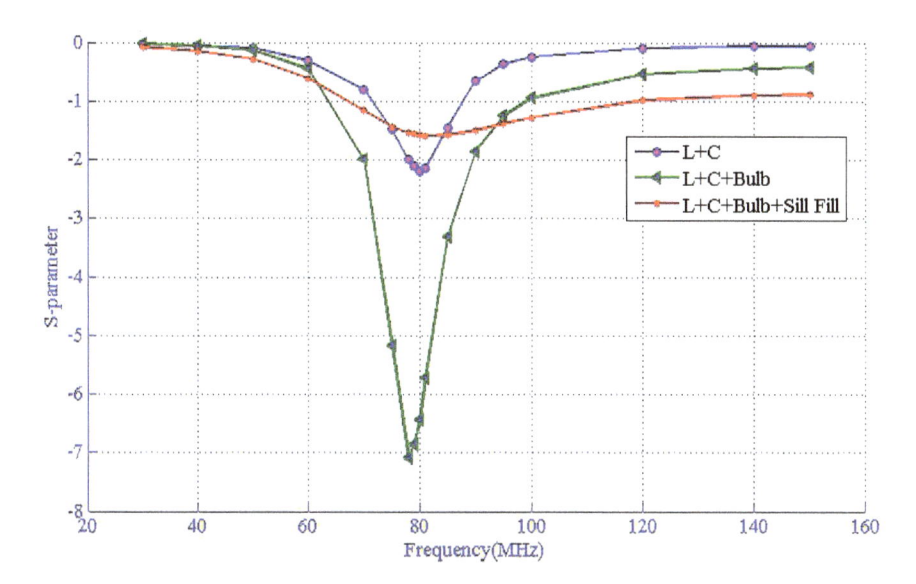

Fig. 3.6 (a) S_{11} = 2dB only with LC resonator (b) S_{11} = 7dB in LC resonator with bulb (c) S_{11} = 1.5dB in resonator with bulb and sill fill epoxy.

industrial development of electronic ballast. These simulators, however, do not incorporate lamp models into their standard libraries and the users are, therefore, faced with the challenge to devise these models separately. Many previously reported fluorescent lamp models [52–54] have approximated the lamp as a power dependent linear resistor or simply a resistor with a cubic voltage-current characteristic. In general, such models are not physically self-consistent and require empirical fittings of data, obtained from the lamp measurements. Their range of applicability is subjected to the conditions under which such data are obtained. Therefore, modelling of Rb bulb is done by measuring the impedances of LC resonator with the bulb and sill fill material, using Smith chart plot. The experimental matching of the impedances of the RF oscillator and the discharge bulb is necessary for the efficient energy coupling.

In this section, the impedance modelling of the LC resonator with bulb load circuit, using ADS software is described. A schematic diagram of the ADS simulation setup source is shown in Fig. 3.7. The ADS is used to extract the parasitic LC components. In a simple

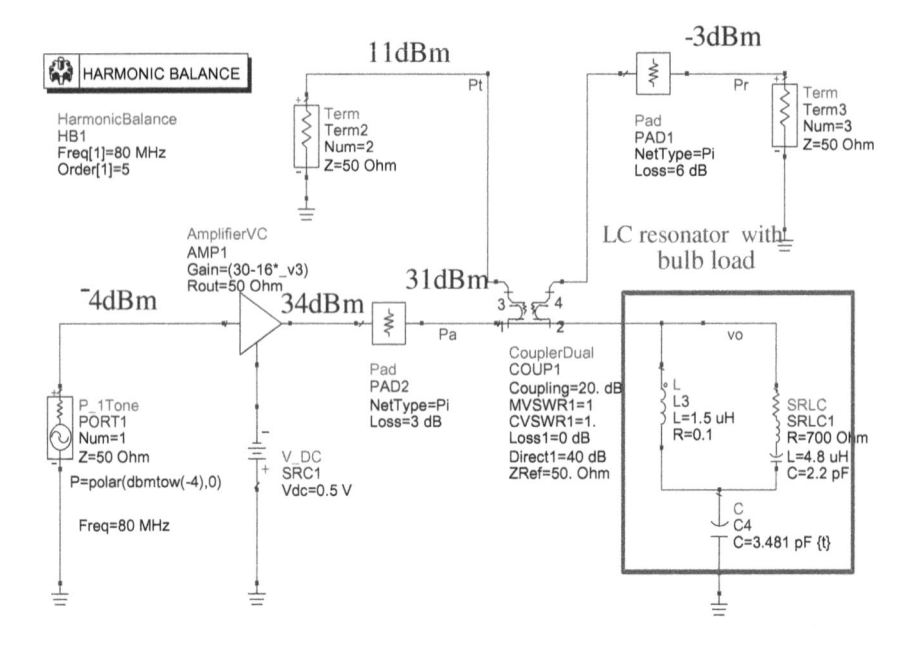

Fig. 3.7 Schematic diagram of ADS simulation setup.

series LCR lumped element bulb model, a capacitance (C) is defined as parasitic capacitance of the plasma bulb. The coil resistance (R) is modelled to show the power dissipation in the plasma through inductive coupling. The value of the R is taken typically ~ 0.1 ohm. If a reasonable estimate of 2.2 pF for the equivalent series capacitance, C and 3.4 pF and for the equivalent parallel capacitance C4 are used in the bulb model, the measured reflection coefficient (S_{11}) is -12dB, when the plasma ignites.

The RF power is supplied to the inductive coil through a variable gain amplifier. The forward terminal of RF amplifier is connected to the input terminal of the directional coupler, and the load terminal connected to the LC resonator stage. A directional coupler measures the incident and reflected powers. This allows the direct measurement of the power delivered to the LC-resonator load, P_{in}, measured forward power minus the measured reflected power,

Studies on Rb Lamp and Driver Oscillator 59

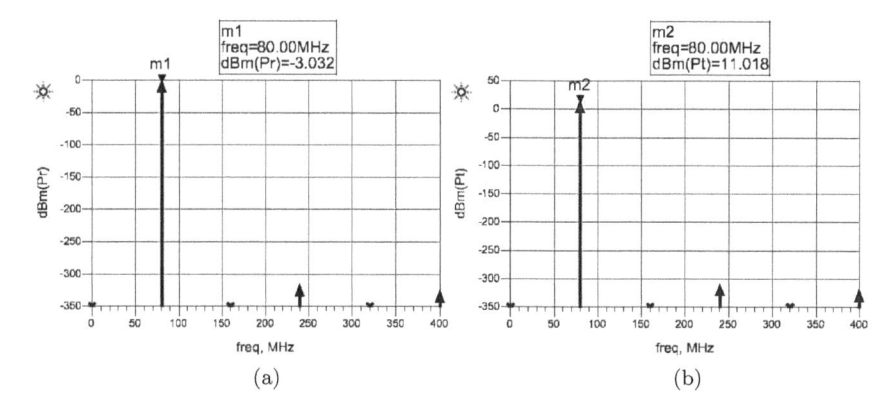

Fig. 3.8 Shows the ADS simulation results of (a) reflected power by the resonant circuit (b) transmitted power by the Rb bulb.

and hence, the voltage, V_0, incident on the LC resonator load. This is necessary for calculating the RF power absorbed across the Rb bulb, using ADS harmonic balance simulation setup. However, the exact RF power and voltage across the Rb bulb is needed for optimizing the drive circuitry, power amplifier or oscillator, and it can be determined by modelling LC-resonator circuit with the bulb load. The harmonic balance simulation results of reflected and transmitted power are shown in Fig. 3.8. The results of the ADS simulation of the equivalent circuit Fig. 3.7, verify the change in absorbed power by the bulb when the plasma exists near the coil. The equivalent circuit model, developed in this work, is used to quantify the reflected and absorbed power by the Rb bulb Fig. 3.8. The ADS analysis of the equivalent circuit model shows that very little amount of power dissipation occurs as inductive heating of the plasma and resistive heating of the coil. While most of the power is inductively-coupled to the plasma through load resistance R. However, a significant fraction of the RF power absorbed by the plasma generator, is dissipated as heat in the coil. Figure 3.9 shows the plasma initiation power in Rb bulb as a function of frequency from 60 MHz to 90 MHz. Although it is possible to ignite the discharge over this entire range, it is clearly easier to start the discharge in the vicinity of 80 MHz.

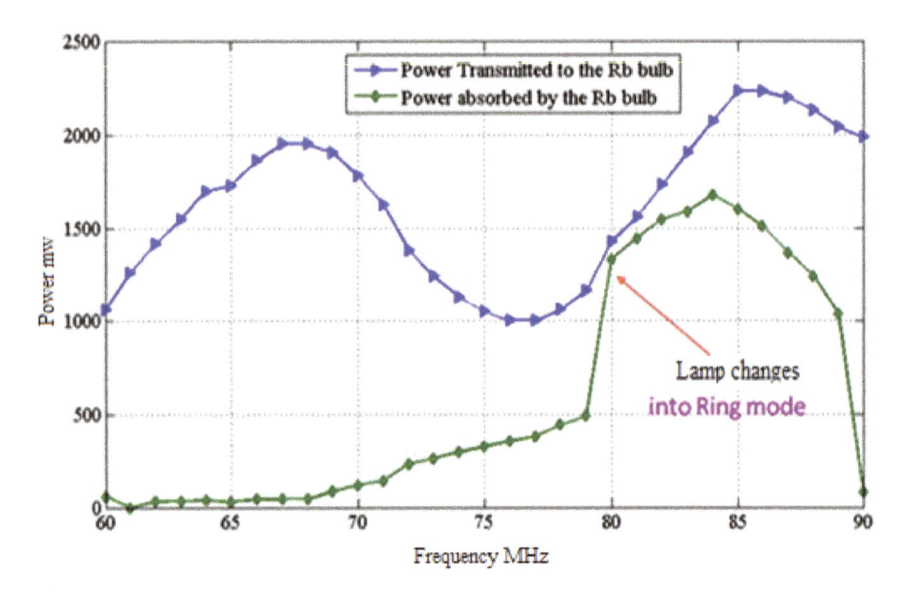

Fig. 3.9 Power absorbed and transmitted by the Rb bulb as a function of operating frequency.

3.3. Experimental test set-up for electrical characterization of Rb lamp

For the purpose of electrical characterization of the Rb lamp, the lamp assembly is shown in the experimental setup, Fig. 3.10. In this experimental test setup of the RF power amplifier and Rb lamp measurement system, the electrical connections are made with 30 W coaxial cable. The input stage of the drive circuit is connected to the RF power source consisting of the Rohde & Schwarz (R&S) signal generator (10 MHz–40 GHz). The Mini circuit ZHL–5W-1, a linear RF amplifier (gain = 45 dB, 5–500 MHz) is used for driving the LC stage: the impedance-matched Rb cell load. The RF signal from R&S signal generator has 50Ω output impedance. Therefore, it is important to carefully impedance match the load using L–C components to a 50Ω source impedance.

This enables high power and voltage amplification across the Rb bulb for getting the required break-down voltage with the lowest input electrical power. Figure 3.10 also shows the RF plasma ignition

Studies on Rb Lamp and Driver Oscillator 61

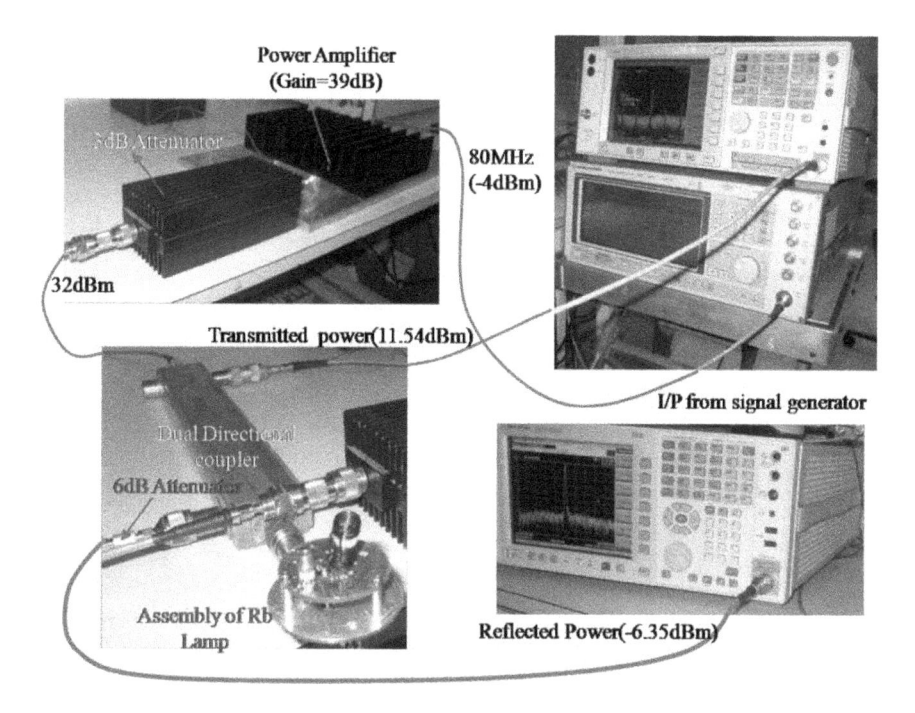

Fig. 3.10 Experimental test setup for Rb lamp measurement system.

circuit connected to the light source, as the impedance-matched load. The two main components in the impedance matching network are (1) a capacitor in parallel to the bulb, to reduce the overall capacitive reactance and (2) an inductor in series with the above capacitors to cancel out the reactive load, thus, forming a purely resistive load. The values and the Equivalent Series Resistance (ESR) of the components are carefully chosen, to maximize the power transfer to the bulb.

3.3.1. *Optimization of RF resonant frequency*

The same experimental set-up as shown in Fig. 3.10 is used to measure the absorbed power by the lamp. The proposed modelling method is used to extract the lamp parameters from the measurements. In the initial stage, a high voltage appears across the unloaded coil, when the RF power source is adjusted to the resonant frequency of the circuit. The RF power source is used for driving this plasma in the Rb bulb, with the characteristic impedance of 50 ohm. The LC driver circuits is built for characterizing the Rb bulb in a frequency range between 30 to 150 MHz. The high-quality coil is tuned to resonate with equally high quality tuneable capacitor at the frequency of interest, which is about 81MHz. The tuneable capacitor is used to cancel the inductive reactance of the circuit, and match the impedance of the plasma to the RF source. The first test excitation frequency is chosen to be 81 MHz for the plasma ignition. Now for the maximum voltage transfer across the bulb, an inductor and a tuneable capacitor (L&C) are added in series with the bulb, as the source, such that the inductive reactance is the complex conjugate of the bulb impedance exactly at 81 MHz, the resonant frequency. Also, as the output terminal of the input stage has a real 50 Ω impedance, an appropriate series resistor needs to be added in the LC stage, to equal the total impedance to a purely real 50 Ω load.

3.3.2. *Power absorption in Rb bulb*

The power required for Rb lamp exciter circuit is optimized. The input and reflected power are measured using a HP 778D dual-directional coupler in the Agilent E4448A PSA series spectrum analyser (3Hz–50 GHz) and Agilent N9030A PXA signal analyser (3 Hz–26.5 GHz). The frequency is gradually increased using Rohde & Schwarz signal generator, to initiate the plasma with the proper discharge mode. Once the plasma is established, the inductive power required thereafter is always considerably lower. The power budget of a lamp exciter circuit is limited. Experimentally, it is observed that the operating mode of the lamp is dependent on the operating frequency and the absorbed power, Fig. 3.11. For the forward RF power, $P_f = 1.5W$, the minimum reflected power occurs at 80 MHz and is very small, and Rb bulb absorbed power is 1.33W (31.54dBm).

In this analysis, one-port S-parameter (S_{11}) is defined as $10\log_{10}$ (P_r/P_f) where P_f is the output transmitted power from amplifier travelling toward the discharge bulb and Pr is the power reflected. The network behaviour of the discharge Rb lamp is plotted in Fig. 3.11. The upper curve in the Fig. 3.11 shows that S_{11} is poor

Studies on Rb Lamp and Driver Oscillator 63

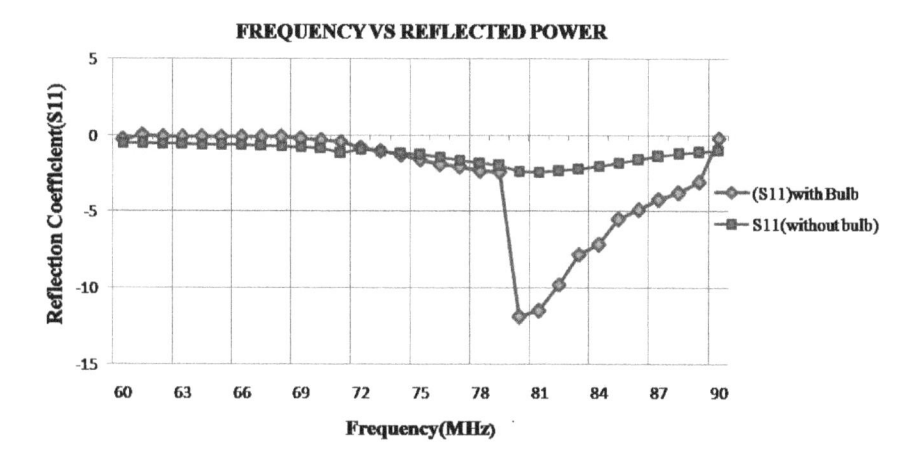

Fig. 3.11 The reflected power vs frequency in Rb bulb.

when Rb bulb is removed from the discharge assembly. A slight dip in S_{11} near 80 MHz indicates the actual resonant frequency of the circuit. The dip in S_{11} in the lower curve occurs when the discharge Rb lamp is ignited. A large negative value of S_{11} indicates that Rb bulb reflects very little power, i.e., the impedance of the discharge bulb is matched to the power amplifier. In the Fig. 3.11, the plasma is shown to ignite in ring mode at $P_f = 1.5\,W$, when the frequency is close to the resonance. However, the power requirement increases rapidly for higher and lower frequencies. The ic solution of fixing the frequency to minimize initiation power is also reasonable, as S_{11} is still $-12\,dB$ at 81 MHz with the plasma on. That is, only 7% of the forward power is reflected from the Rb bulb. Therefore, only 1.5W of RF power is required by the RF oscillator to run the Rb bulb, which is far less than the power needed with RF amplifier to initiate the bulb ignition.

3.3.3. Characterization of Rb lamp mode vs. frequency

The lamp mode is also changed from dim pink to ring mode at 80MHz when the bulb impedance is matched to the output impedance of power amplifier. From 80MHz to 87MHz operating frequency, the bulb is operated in the ring mode. The red mode appears at 88 MHz and is converted to dark red at 90 MHz. Above the 90 MHz the bulb intensity is drastically reduced and appears as dim pink. The entire experimental method has been simulated using ADS simulation software with a source frequency of 80 MHz. The observed transmitted and reflected power is shown in Fig. 3.9. The shift in the resonant frequency of the Rb bulb is observed when the discharge is ignited, which can be understood by considering the bulb capacitance change. The frequency shift caused by the capacitance change of the ignited Rb bulb may affect the design of the RF power oscillator. The variations of bulb mode with frequency are show in Fig. 3.9.

3.4. Rb lamp driver circuit design aspects

For driving the electrode-less Rb lamp, the self-excited oscillator is more often assembled with the inclusion of an oscillatory circuit in the grounded base clap oscillator configuration. The goal of this research and development is to develop an efficient Clapp oscillator based Rb plasma light source, shown in operation in Fig. 3.1. This utilizes the inherent advantages of Rb discharge lamp. The entire assembly, including the ballast, is contained in a copper housing, as shown in Fig. 3.5. The operating RF frequency in the range of 60–100MHz, is generated by a tunable high frequency RF oscillator with the maximum power of 6W. The main section of this Rb lamp assembly is a spherical shaped Rb bulb of 10mm diameter, placed inside an inductor coil, having inside-out turn configuration.

Figure 3.12 shows a common-base Clapp oscillator using a NPN transistor as the amplifying device. In the common-base configuration, there is no phase difference between the signal at the collector and the emitter signal. A Clapp circuit is preferred to a Colpitts circuit for the Rb lamp exciter. In a Colpitts oscillator, the voltage divider contains the variable capacitor (either C4 or C5). This causes the feedback voltage to vary, which at times make Colpitts circuit less likely to achieve oscillation over the desired frequency range. This problem is avoided in the Clapp circuit, by using fixed capacitors in the voltage divider and a variable capacitor in series with

Studies on Rb Lamp and Driver Oscillator 65

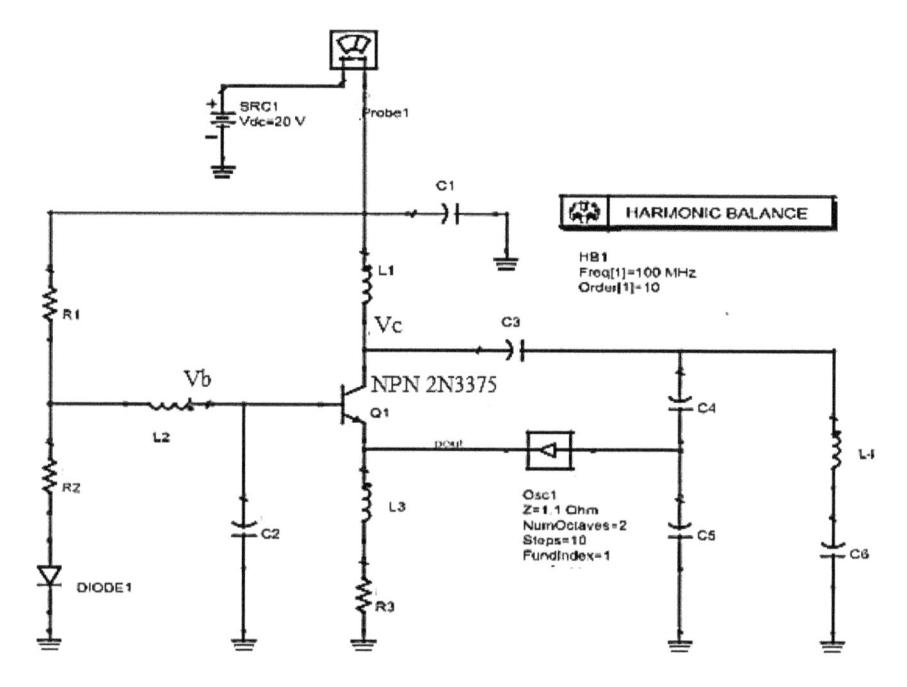

Fig. 3.12 Schematic of common-base Clapp oscillator circuit.

the inductor. The Clapp oscillator is in a way an improved Colpitts oscillator with an extra capacitor labelled as C6 in series with the coil, as seen in Fig. 3.12. The function of C6 is to reduce the effects of junction capacitance on the operating frequency. Resistors R1 and R2 provide the usual stabilizing DC bias for the transistor in the normal manner while the capacitor C3 acts as DC-blocking capacitor. The capacitor C6 is always much lower in value than either C4 or C5, so it becomes the dominant capacitor in any frequency calculation. C4 and C5 provide the phase shift needed for the regenerative feedback.

In view of the critical role of Rb lamp, the selection of the transistor is an important step in designing the lamp exciter circuit. It depends on the factors, such as output power, oscillating frequency and supply voltage. In this scheme, a high power RF BJT (2N3375) is used as an amplifier. The circuit is simulated on Agilent Advance Design System circuit simulator. The 2N3375 BJT from Microsemi

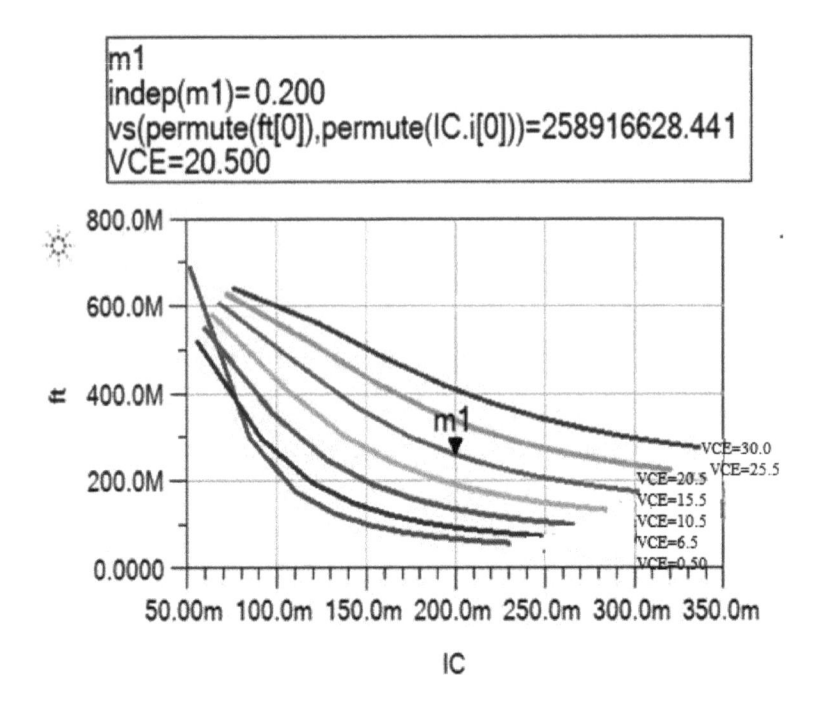

```
m1
indep(m1)=0.200
vs(permute(ft[0]),permute(IC.i[0]))=258916628.441
VCE=20.500
```

Fig. 3.13 Plot of f_T vs. I_C with different value of VCE.

has f_T of 259 MHz for 200 mA collector current, roughly 2.5 times above our desired operating frequency (Fig. 3.13). This is sufficient to achieve a reasonable loop gain to permit oscillation [55] but is low enough to prevent the transistor from strong compression and to reduce oscillations at higher harmonics. Therefore, a few dB of compression is desirable to ensure the initiation of oscillation over a temperature variation and various tolerances.

The bias network, typically includes an inductor (L1) high impedance line at the resonant frequency (f_o), connected through a bypass capacitor to the ground. So that no power at resonant and other harmonic frequencies leaks into the DC bias. The operation of a bias inductor appears as an open circuit to a transmission line. The capacitor acts as a short circuit at the resonant frequency. The bias tee for collector and base terminal is shown in Fig. 3.14.

For the frequency up to 2GHz, a well-designed lumped element bias tee circuits can be used for good results. The simple L-C circuit

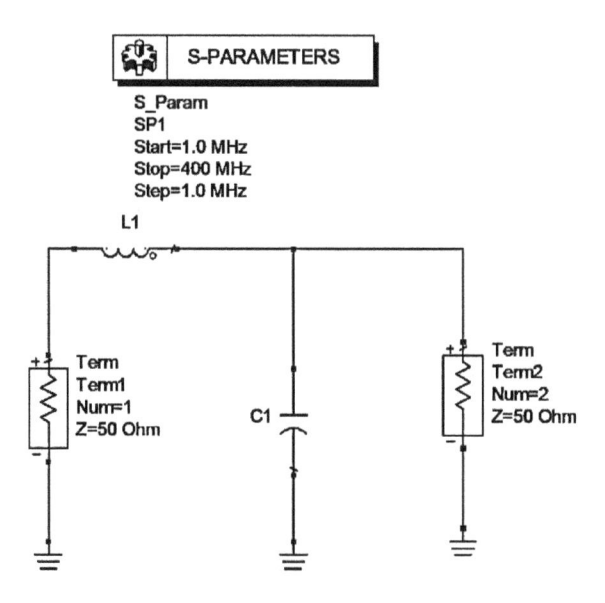

Fig. 3.14 Schematic of bias tee.

has adequate insertion loss in this operating frequency band as shown in Fig. 3.15.

3.5. Configuration simulation and development of Rb lamp exciter

During the design process, both linear and non-linear analysis of the oscillator is performed for their respective advantages. The linear analysis is based on small-signal, linear S-parameters. It only guarantees initial starting condition and gives an approximate value of the oscillation frequency. The non-linear analysis is based on the true large signal conditions. It can be used to determine the frequency of the oscillation precisely under steady state, as well as several crucial non-linear parameters, such as output power, phase noise, pushing and pulling, etc. [56].

The active device in linear and non-linear model is 2N3375, and the bias point is set at $Vc = 20.5\,V$ and $Ic = 200\,mA$. The linear model has the same topology as the non-linear model, except the

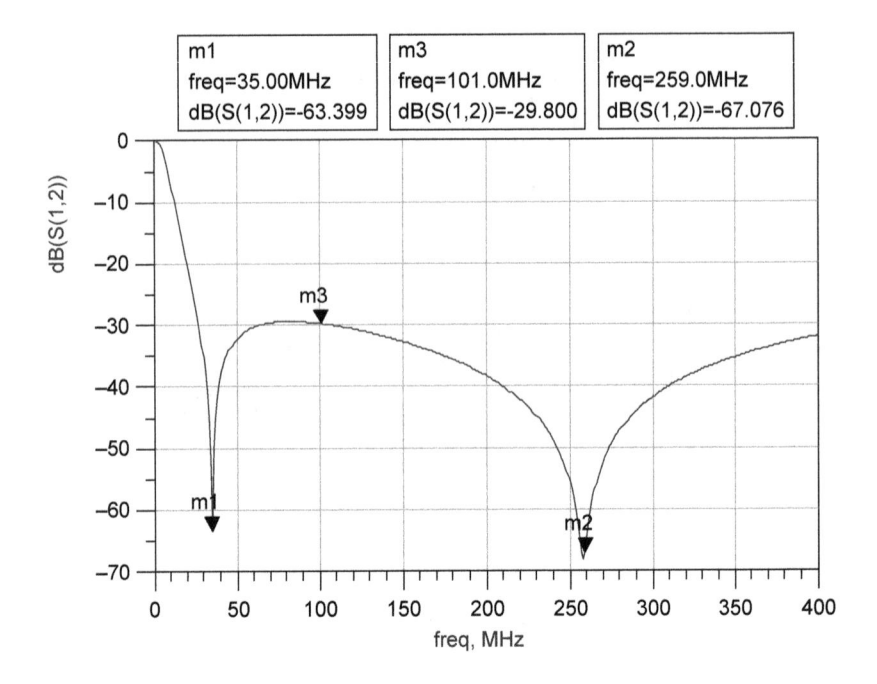

68   *Rubidium Atomic Clock: The Workhorse of Satellite Navigation*

Fig. 3.15   Response of bias tee.

following differences:

(1) In ADS simulator "Osctest" in linear model is replaced with "Oscport" in non-linear model.
(2) S-parameter simulator is used in linear model and in the non-linear model harmonic balance is utilized.

3.5.1. *Linear circuit simulation*

The linear oscillator circuit is cascaded with the amplifier-resonator. The open loop bode response (magnitude and phase of S_{21}) is used to determine the gain margin (small-signal open loop gain at the phase-zero crossing) in the oscillator design by breaking the feedback loop and terminating it with 50 ohm impedance. The amplitude of the forward scattering parameter S_{21} and phase of S_{21} of the cascade are plotted in Figs. 3.16 and 3.17 respectively.

The phase of S_{21}, plotted in Fig. 3.17, explains that ϕ_0 (zero crossing) occurs at 100 MHz with the maximum phase slope.

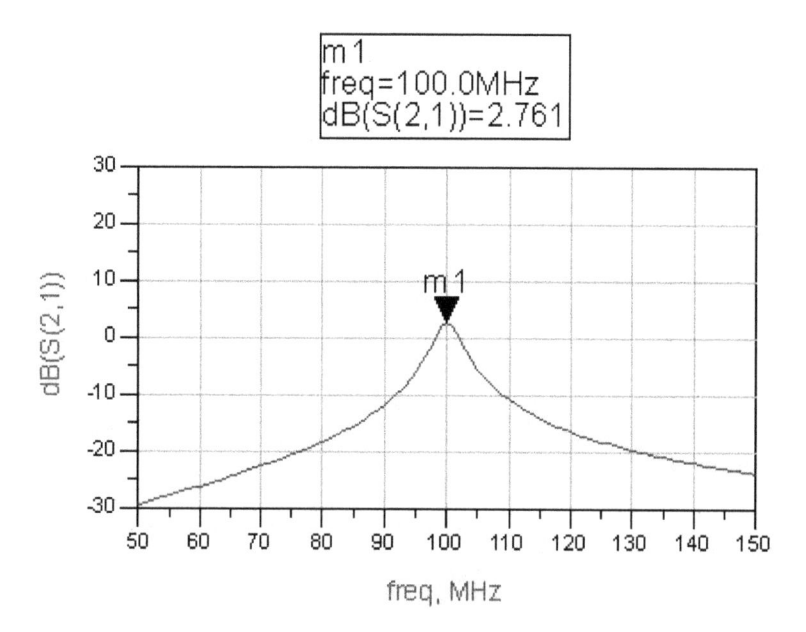

Fig. 3.16 Forward S-parameter S_{21} amplitude for the amplifier-resonator cascade. The gain peaks at 100 MHz at about 2.7 dB.

3.5.2. Non-linear circuit simulation

The linear techniques cannot predict the oscillator's fundamental output level, the harmonic content of the output spectrum, internal voltage and current waveforms. As the initial signal builds in an oscillator, the non-linear action absorbs the open loop small signal gain margin to establish a steady-state operating point. The non-linear behaviour of the oscillator, the harmonic content of the output spectrum, or internal voltage and current waveform results are simulated by the harmonic balance (HB) simulation engine of the ADS software, by Agilent Technologies. The HB simulation of the oscillating circuit is performed by closing the loop. The harmonic balance (HB) simulator [57] and the ADS test element "OscPort" are used for non-linear oscillator circuit analysis, noise analysis and its optimization. In the HB simulation, the "OscPort" is used instead of the "OscTest" element. The "OscPort" is a special element used in HB analysis of an oscillator where the simulator must find solution for both frequency spectral. It assesses the oscillator's feedback loop

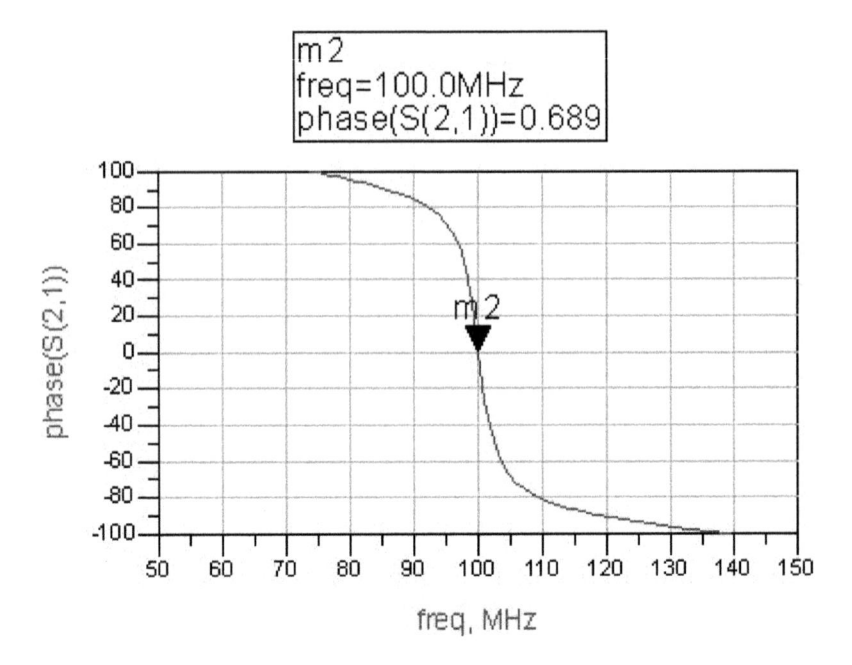

70   Rubidium Atomic Clock: The Workhorse of Satellite Navigation

Fig. 3.17 Linear S-parameter (S_{21}) phase for the amplifier-resonator cascade. The phase is zero at 100MHz.

without breaking it or altering the circuit characteristics. The HB simulator and the "OscPort" component automatically determine the operating characteristics.

3.5.3. Oscillator layout design and development

The complete layout and the actual fabricated oscillator circuit are shown in Fig. 3.18. The oscillator circuit is fabricated on the 1.6 mm FR-4 substrate with the dielectric constant of 4.5 and based on a laminate board using SMT devices. When integrated with the lamp assembly as described in the next chapter in more detail, the PCB is extended to reach the transistor. The ground plane on the backside of the PCB is used to permit mounting on several structures, such as the brass test fixture, PCB holder etc shown in Fig. 3.18. This allows the Rb bulb assembly with inductive coil and thermal control circuitry to be integrated to the lamp assembly. The oscillator has two ports, one for input DC bias and other for capacitive

Fig. 3.18 The complete fabricated oscillator PCB with PCB holder, Rb bulb housing and assembled thermal heater.

feedback. The power supply noise on the bias line is filtered, with two capacitors (typically 1 uF and 820 pF), to ground, and a RF blocking is provided by a 1uH inductor. The test circuit is milled from a single piece of brass to allow a solid mount, a good heat sink, and also a brass lid for electrically shielding the bulb assembly during RF power measurements. All components are space qualified to provide a repeatable and inexpensive design. The layout is compact and designed in a circular 4 layer PCB. The largest component is the power transistor which sets a size limit for the oscillator circuit, since other components are chosen as size 0402 (40 mils by 20 mils) and are surface-mountable for ease of placement.

3.5.4. *Studies on Rb lamp ignition voltage vs operating frequency*

The experimental investigation is done on the dependence of the HF voltage across the inductor for the ignition of the discharge in

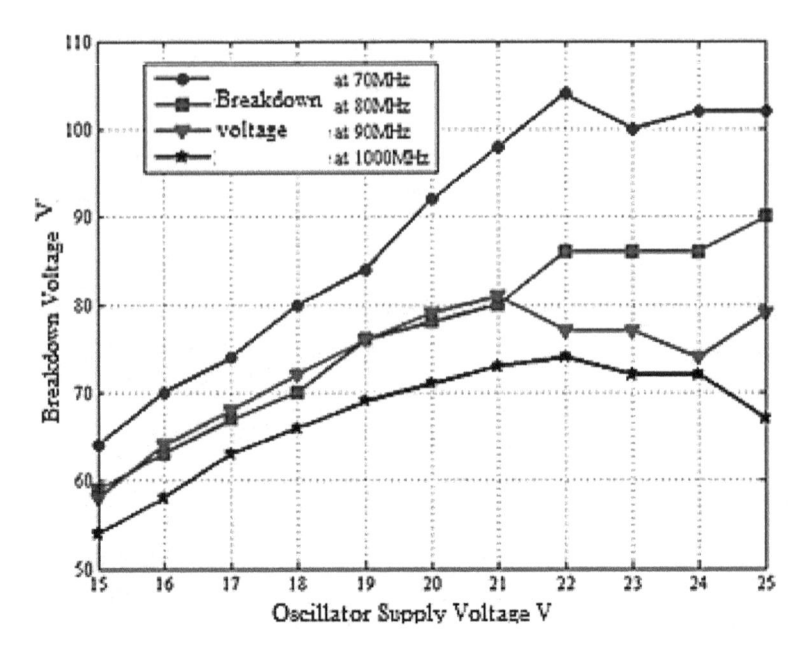

72 Rubidium Atomic Clock: The Workhorse of Satellite Navigation

Fig. 3.19 Dependence of breakdown voltage on the frequency of the exciter field.

the lamp. The excitation of the discharge as a function of frequency (between 70–100 MHz) is especially investigated. In the passive Rb atomic clock, the frequency of RF field for excitation of ^{87}Rb lamp is generally several tens of MHz. For ^{87}Rb lamp with a diameter of 10mm; we study the frequency range up to 100 MHz. The results of the lamp breakdown voltage for Rb bulb of diameter 10mm, at frequencies 70 MHz, 80 MHz, 90 MHz and 100 MHz are plotted in Fig. 3.19. It can be seen that for each working frequency, there is a working point where the breakdown field achieves the minimum. Along with increasing the frequency of RF field, the minimum electric field required for breakdown of Xenon and Rb decreases. The explanation is, the higher the frequency the more the energy electrons are obtained for the ionization, leading to breakdown of gases to occur more easily even at a lower electric field, at the same gas pressure. In practice, we require the breakdown electric field to be as low as possible. It can be seen that the minimum excitation voltage for breakdown is reduced from 100V to 72V when the frequency of the RF field increases from 70 MHz to 100 MHz.

3.5.5. Voltage-current characteristics of the Rb lamp oscillator

The voltage-current characteristics of the discharge electrode-less (EL) Rb lamp operating at 80 MHz frequency are experimentally observed and are shown in Fig. 3.20. With the increase in the voltage, an increase in current, approximately, proportional to the voltage is observed, until the ignition is initiated. When power is switched on, a stable pink mode-discharge appears and an insignificant change of the parameters is observed. In the case of transition from pink mode to ring mode, a significant jump of the current, caused by the essential change of load of the oscillatory circuit takes place. The ignition current through the circuit for ring mode-discharge can increase or decrease depending on the relation between the internal resistance

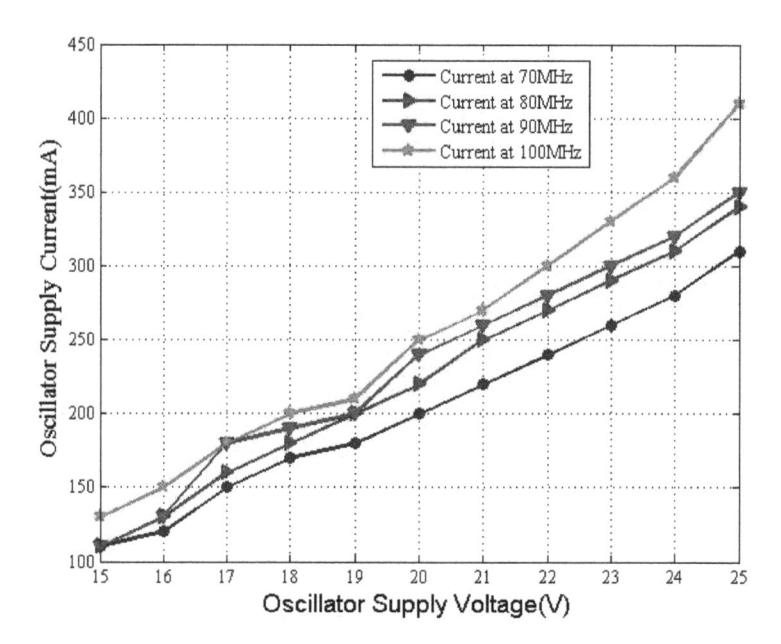

Fig. 3.20 Voltage-current characteristics of the electrode-less (EL) Rb discharge lamp.

and the resistance induced by the discharge. Therefore, a constant voltage is applied to stabilize a desired mode. It is clear from the plot in Fig. 3.20 that the current varies almost linearly with the supply voltage at an operating frequency of 80 MHz. A mode stabilization is observed at a supply voltage of 20V. Therefore, a constant 20 V power source is finalized for the lamp operating frequency at 80 MHz.

3.5.6. *Output power and efficiency*

For the optimal performance, the Rb lamp should be driven at a specific RF oscillation frequency; with a minimal of DC power and absorbed power. The oscillator operates at 20 Volt and a 200 mA current. For the RF power measurements, a 10 dB high power attenuator is attached at the RF output of the test board. The RF power is coupled from the lamp with a capacitor divider and is measured in the spectrum analyser. Figure 3.21 shows the output power spectrum

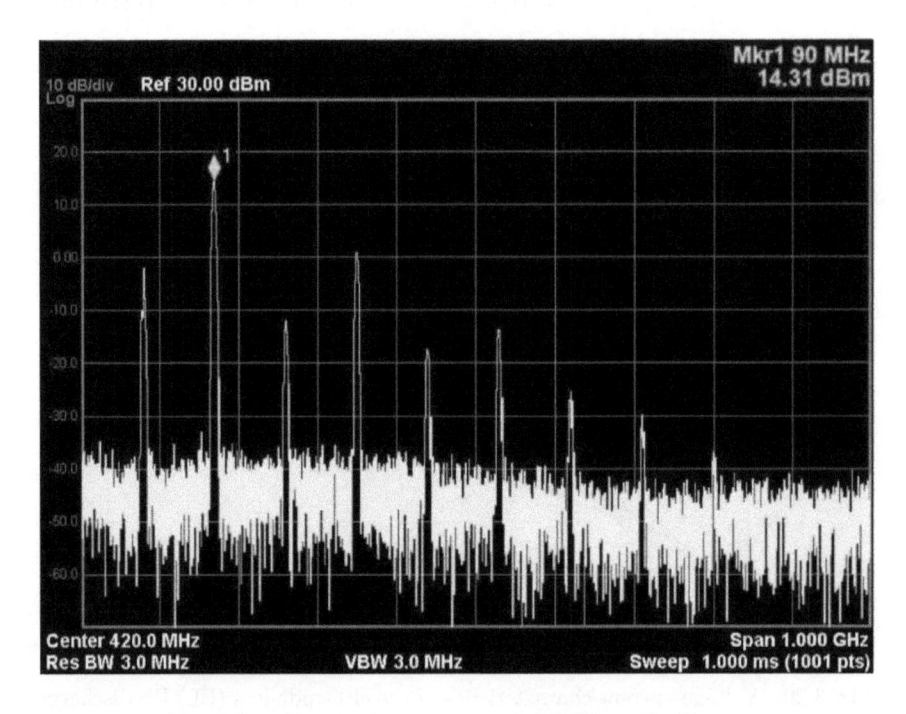

Fig. 3.21 Output power spectrum of Clapp Oscillator.

in the frequency domain of PSA (performance spectrum analyser) E4440 at the steady state oscillation condition.

3.6. Spectral profile of Rb lamp and operating temperature vs modes

3.6.1. Introduction

The performance of the atomic clock depends on the spectral stability and output of the RF discharge lamp, that is very sensitive to the temperature. There are many shapes used in electrode-less lamps [58–59]. Each shape offers a different optical or the efficiency advantage. Some lamps are spherical in shape with the inductive coupling coil outside the lamp, protected by a brass cylindrical cavity. Some spherical lamps have the coil surrounding the bulb. We examine the spherical shaped lamp which uses two inductive coupling coils connected in parallel. The physics of this lamp has been discussed in recent papers [60–63]. The results can be applied to other electrode-less lamp shapes. Different groups [64–70] have carried out detailed studies on the possibilities of utilizing such light sources.

The spectral lines and voltage-current characteristics of the lamps filled with different buffer gasses, and the effect the temperature on the spectral lines are studied in detail. These studies, as well as further research [70, 71] make it possible to develop spectral alkali metal light sources and a number of other elements for the purpose of effective optical pumping. It is found that such a lamp radiates extremely narrow spectral lines, suitable for various applications specially, atomic clocks. The light emission characteristic of the discharge lamp can change slowly over time, and may affect the long-term stability of the atomic clocks [72].

The Rb light source is a glass bulb approximately 10 mm in diameter filled with [87]Rb isotope or natural Rb and a buffer gas Xenon. The proper operation of the Rb frequency standard using an alkali vapour lamp conjugated with the power oscillator can reliably start the lamp and maintain a constant lamp output under varying environmental conditions. Although Rb lamps are very reliable, the exact time of the ignition after RF turns on is not controlled. The RF excitation applied to the lamp should be sufficient to sustain the plasma discharge of the Rb vapour and buffer gas, however, other factors also influence the starting. The starting may be delayed by the presence of an invisible, conductive Rb film on the internal surfaces of the lamp. This film can reduce the RF excitation within the lamp. The heating of the lamp appears to drive off the film, enabling the lamp to start. The Rb lamp is excited by an RF power oscillator delivering a power of the order of 3–4 watt at a frequency, 60MHz–100 MHz. The temperature, mode and RF power variations in the bulb can change the lamp output, both in intensity and spectral distribution [73].

In this section we discuss the behaviors of spectral characteristics of the Rb lamp with temperature. The spectral profile of D1 line (794.8nm) and D2 line (780nm) are observed over the temperature range from 23°C to 110°C. The Rb lamp exciter power and temperature of Rb bulb are very important parameters in controlling the performance of the Rb lamp. It is observed that at temperatures beyond 110°C, the lamp mode changes from the ring to red mode, resulting in abnormal broadening of emission lines and self-reversal. The relation of the radiation intensity to the mode of the lamp is established, and a limiting upper intensity of radiation of lamps with a particular diameter filled with a mixture of Rb vapour and Xenon is stated. The discharge lamp operated in two distinct spectral modes, which are referred to as the ring and red modes [74–76]. The temperature dependence of these two modes and optical intensity variation with time are important aspects of the lamp. The lamp intensity stability and the optimum temperature of operation are monitored.

3.6.2. *Experimental set-up*

As shown in Fig. 3.22, the Rb bulb, under observation, is a spherical glass bulb of diameter 10mm. It contains natural Rb atoms and Xenon gas at 2 ± 0.2 torr, is supplied with 55–110 MHz RF power and can be heated from room temperature to 110°C.

The Rb lamp is bonded using RTV to a copper lamp holder, an integral part of the PCB, that is also used as a cover for the lamp. The

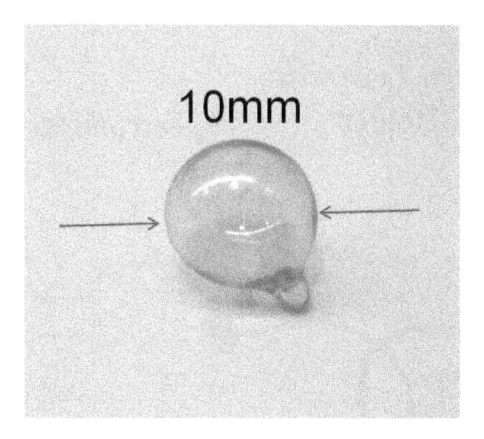

Fig. 3.22 Photo of Rb bulb with 10mm diameter.

assembly consists of a copper oven with an open window. Increasing the lamp oven power to improve the warm up time has limits. The lamp exciter also helps in heating the lamp. The power dissipated in the lamp exciter electronics conduct through the bolted interface and travels to the lamp oven. The lamp oven heaters are made using two layers of resistive foil. Therefore, the current flows in each heater in opposite direction that cancels the residual magnetic field. The foil layers are insulated from each other with layers of kapton and adhesive. The construction of the heater and the adhesive interface to the copper oven shell create a thermal resistance. To warm the glass shell of the lamp, the heat must conduct through the copper holder of the lamp holder. The lamp holder is modified to include a groove at the base. A twisted pair of fine resistance wire is wrapped into the groove and bonded with a conductive epoxy. The temperature set point for the groove is a few degrees cooler than the lamp operating temperature. The tip or the groove of the bulb acts as a Rb metal reservoir, regulating the Rb vapour pressure in the bulb. This helps in ensuring the long life of the Rb lamp and uniform intensity over the period of its operation.

As there are Xe/Kr atoms in the Rb bulb, the spectra measured are divided into two parts. One is the spectra of Rb atoms which is investigated in this work. The other is the spectra of Xe/Kr

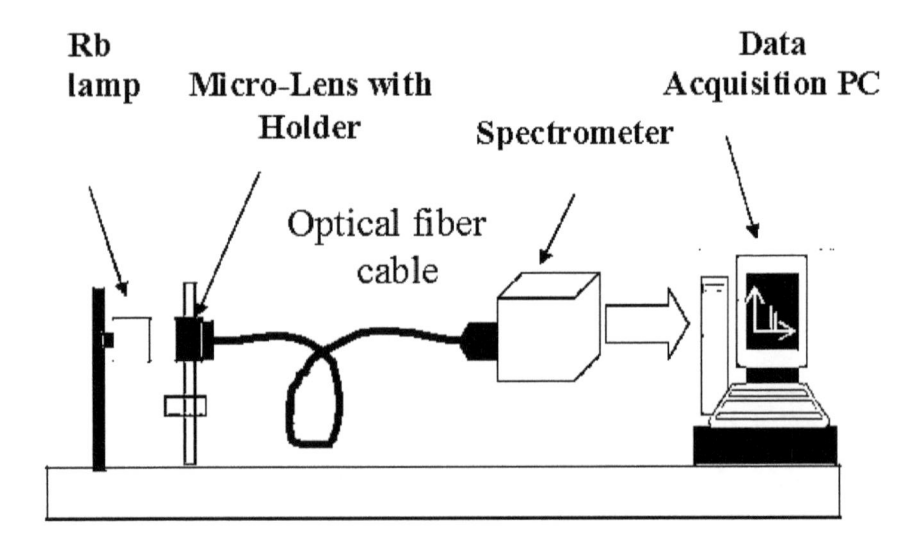

Fig. 3.23   Experimental test setup for studying the spectrum of Rb lamp.

atoms. The emitted spectrum from Rb lamp is measured by optical spectrometer (USB2000+ spectrometer produced by Ocean Optics Company in USA with a resolution of 1.5 nm) and an experimental assembly is shown in Fig. 3.23. The energy levels of Rb atoms, that we need to measure are located in the region of 350–1110 nm. For simplicity, we do not consider the working mode when the intensity of fluorescence spectral lines of Rb atoms is measured. The light from the Rb lamp is incident on the micro-lens which focuses the beam into the optical fibre. The fibre guides the beam to the grating based spectrometer, having a range of 339.48–1022.05 nm and step size of 0.28 nm. The spectrum is acquired by a 2048 pixel linear CCD detector, and plotted on data acquisition PC linked to the spectrometer.

3.6.3. Spectroscopy of Rb lamp

The radiation intensity of lamp, with alkali metal vapour, depends strongly on the temperature of the lamp. The spectrum is observed at different thermostat temperatures ranging from 65°C to 120°C, by heating the Rb lamp using bifilar Kapton heater and controlled with a PID temperature controller. The temperature is maintained

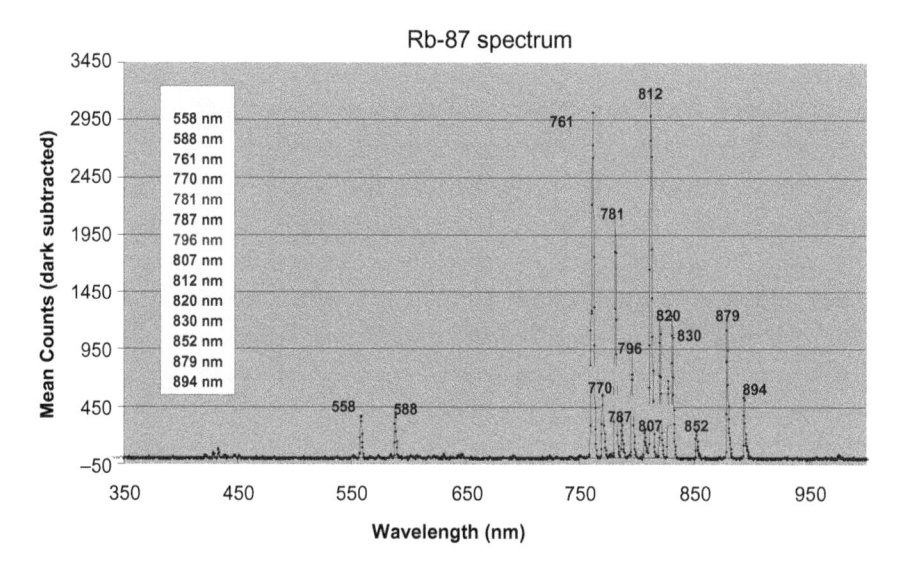

Fig. 3.24 The spectrum of Rb lamp along with Xenon lines.

with a stability of ± 0.1°C. The bulb temperature is measured using the Pt-100 temperature sensor. The power consumed by the self-excited oscillator is stabilized with an accuracy of 1%. The variation of the intensity of the Rb spectral lines with the temperature of the lamp and the power of the discharge, is measured with an optical spectrometer. The shape of the spectrum profile of the lamp at the ambient temperature of 25°C includes that of the buffer gas Xenon and the D1 and D2 lines of ^{87}Rb, shown in Fig. 3.24. The intensity of different spectral lines of Rb atoms varies widely, 812 nm is the strongest, 761 nm and 781 nm are ranked second and third respectively. The spectral line 558 nm next to 588 nm is the weakest as observed in our measurements. The spectral stability of these lines are studied, at normal room temperature versus the time, by monitoring the mean count of these lines, over a period of nearly three and a half hours. The spectrum is recorded at hourly intervals. However, for Xenon, the magnitude is insignificant ($\approx$ 14 counts/1–2%) compared to D2 ($\approx$ 275 counts/16%). The variations of mean count at normal environment temperature versus time are due to the temperature and RF power fluctuations. These lamps, although self-heated by the RF exiting power, need to be temperature controlled. The RF excitation heats the lamp, due to the plasma discharge and dielectric heating. The glass envelope of the lamp then heats Rb metal to generate Rb vapour. After electrons and ions of the lamp containing Rb and Xe are accelerated by the high frequency electric field, their energy increases [77]. These electrons and ions with high energy collide with Xe and as a result, more electrons and ions are produced. These electrons and ions with high energy excite Xe atoms to higher energy levels. When these Xe atoms in the excited states collide with Rb atoms, energy can be transferred from Xe atoms to Rb atoms to excite them.

At the room temperature, the vapour pressure of Rb is very small ($p \sim 10^{-6}$ mbar). This corresponds to too little ionization, in the bulbs with diameter in a few mm-scale, for an electrical breakdown and eventual discharge emission. In order to have a higher ionization efficiency [78], the temperature of the bulb is increased to more than 100°C. By adding a buffer gas, the Rb vapour pressure can be increased leading to a higher ionization efficiency, without increase in input power or bulb size. Figs. 3.25(a)–(d) show the dependence of the intensity of the D1, D2 and Xenon inert gas line on the temperature. From Fig. 3.25(c) it appears that the buffer gas Xenon is essential for starting the discharge and maintaining a stable state after equilibrium is reached. Xenon is the best as the starter gas due to its lower ionization potential. The maximum intensity of Rb on the D1 & D2 lines is realized at a temperature of 100°C. On further increase in temperature, the intensity of these Rb lines show downward trends. The actual intensity variation of D1, D2 and Xenon lines with temperature are more visible by separately observing three spectral lines and their intensity ratios at different temperatures.

From Figs. 3.26(a)–(b), it is clear that the intensity of spectral lines and the intensity ratio grow with increasing temperature and reaches the peak at 65°C. A further increase in temperature results in rapid intensity drop up to 85°C. The intensity becomes almost constant in a range of the temperature from 85°C–90°C, then again it reaches its peak at 100°C. After achieving the maximum intensity at 100°C it decreases quickly. The observed intensity reduces for

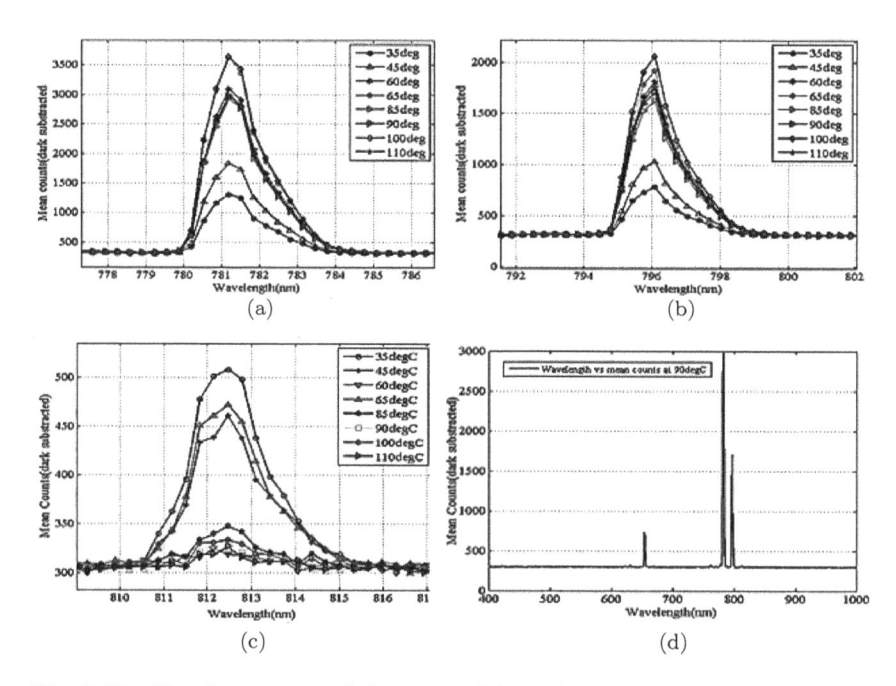

Fig. 3.25 The dependence of the spectral intensity variation the temperature (a) D1 line (b) D2 line (c) Xenon inert gas line (d) Spectrum of D1 & D2 line at 90°C.

temperatures between 65°C to 85°C and also above 100°C because of the intensity drop of the inert gas lines and a significant reabsorption of spectral lines [79].

3.6.4. Operating mode of Rb lamp

The operation mode of the lamp and its design are conceptualized with the selection of the bulb dimensions [80]. To increase the longevity and reliability of the lamp and to reduce the power consumed by the thermostats, it is expedient to select a mode with the smallest possible power for the discharge. This is feasible with the higher gas pressure, and with the lower temperature of the lamp. Thus the thermostat temperature should be chosen to provide the highest intensity of radiation for a given discharge power.

The Lamp dimensions can be reduced with a little increase in the oscillator power and thermostat temperature. For the E-discharge

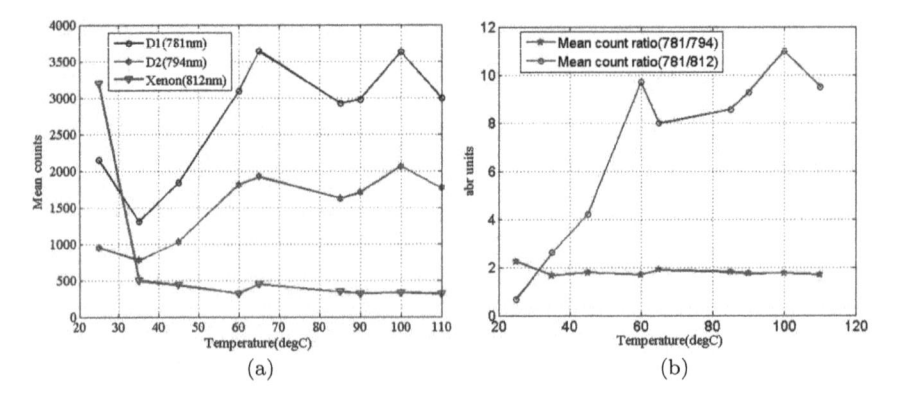

Fig. 3.26 (a) The intensity of spectral lines D1, D2 and Xenon (b) Mean count ratio of D1 line with respect to D2 and Xenon line.

mode, a reduction in size of the spherical bulb is possible down to 8mm diameter. Further reduction in lamp dimensions leads to lower reliability. The Rb aging is increased at higher temperature and the control of the Rb metal depletion rate may become difficult. Therefore, we choose the Rb bulb of dia 10 mm for the qualification testing.

The lamp can operate in three spectral modes. The ring mode, the red mode and the weak mode [81]. The modes of the Rb bulb depend on the absorbed RF power, oscillator frequency, and temperature of the bulb. On increasing the temperature of the lamp, the radiation intensity of the resonance line of Rb initially varies in proportion to the concentration of Rb atoms in the ground state. The increase in the intensity is less than the growth of the number density. The reason for this is the re-absorption of the radiation, which may be estimated by means of the intensity of the Rb D_1 and D_2 lines. At the same time the spectral lines becomes self-reversed [82]. A visual examination of the lamp and the monitoring its temperature show that in the temperature range of 65°C to 75°C, the mode of operation changes, somewhat whitish colour at the centre, typical of the noble buffer gas and an overall pinkish-violet ring, a characteristic of Rb. At the temperature above 110°C it changes from ring to red mode [83]. Under the nominal conditions, the transition between these two modes is "abrupt" for a lamp temperature above 110°C. The mode transition temperature depends on the RF power. When changing the RF power from low to high; the white, the ring and the red modes appear in a sequence. Irrespective of the mode in which the lamp operates, the population ratio between the excited states of Rb atoms is almost constant. Additionally, we find that in the ring mode both alkali and noble gas spectral lines are present in the lamp's radiation. While in the red mode only alkali lines appear. Though the discharge lamp of the Rb clock is conceptually simple in design [78, 81, 82, 84 and 85], its operation can exhibit interesting behaviour. Of particular importance is the "ring-mode" to "red mode transition". From Fig. 3.26, it is interesting to see that the operating temperature lies between 85°C–90°C, where the spectral lines are clearly not self-reversed. Figure 3.27(a) shows the white mode of the Rb lamp. At low lamp operating temperatures, the lamp appears white in its centre and pink/red near the walls as illustrated in Fig. 3.27(b). In this ring mode, the amount of the Rb light emitted by the lamp increases with temperature and so does the Rb clock signal. At higher lamp temperatures, the discharge becomes reddish throughout. In the red mode, the clock signal is a decreasing function of lamp temperature and the Xe emission "turns off". Figure 3.27(c) shows the red mode excitation of the Rb lamp. The stability of the lamp light output is determined by measuring the output current fluctuation of the photodiode, over a period of nearly two and a half hours at a fixed temperature of 87°C.

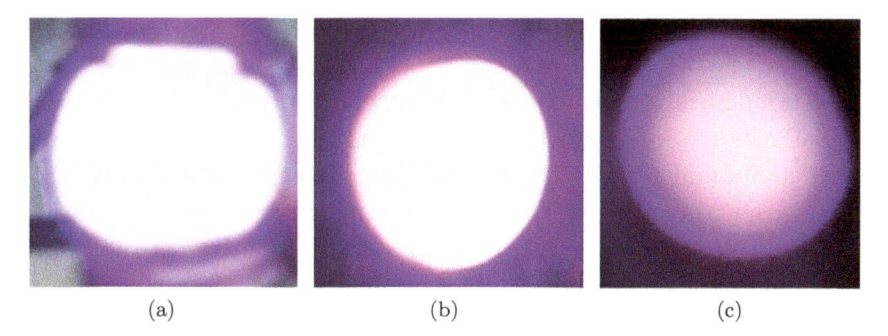

Fig. 3.27 Three different transition mode of Rb bulb (a) White mode (b) Ring mode (c) Red mode.

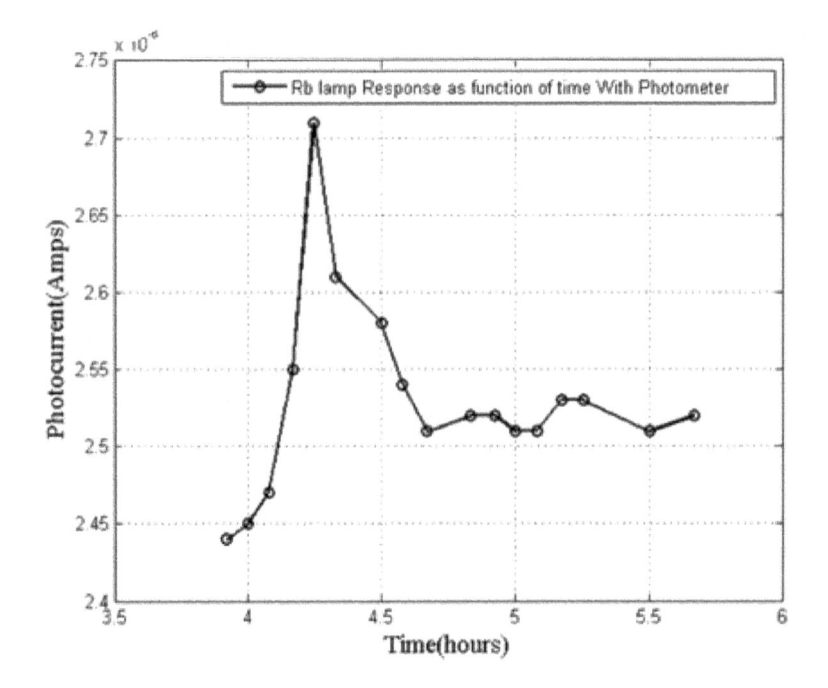

Fig. 3.28 The Rb lamp output (Photodiode current) as a function of time.

Some transients in the photodiode current are also observed. It is due to the power supply current variation of oscillator circuit, which may affect the lamp stability. Figure 3.28 shows the photodiode output current stability of the Rb lamp as a function of time.

Summary

The temperature dependence of the spectral profile of the ^{87}Rb D1 and D2 lines is observed by optical spectrometer. It can be inferred that at around 65–100°C, the isolation of the D1 and D2 lines of ^{87}Rb from stray lines of Xenon is possible. However, as temperature plays a very important role in the proper functioning of the lamp, it is required to incorporate an additional temperature controller.

From the study of spectral stability of these spectral lines with time, it is revealed that the intensities of D1 and D2 lines are sensitive to the changes in temperature and absorbed RF power by the bulb. Under the normal environment, variation of the temperature with time causes 15–16% variation of means counts/intensity. Therefore, the temperature stabilization of the lamp is addressed with the careful experimental probing. We also examine the ring-mode to red mode transition in alkali RF-discharge lamps. From the above extensive discussions, it is established that the temperature and RF power play crucial role in the mode transition.

Chapter 4

Thermal Analysis of Rb Physics Package Components

> In the satellite navigation, the heat dissipation is a very critical aspect. Radiation and conduction are the only means of heat management. The Physics package of the onboard Rb atomic clock/s has the lamp, heaters and base plate which need thermal stabilization and equilibrium. We report in this section thermal analysis of Rb Physics package in details taking care of every minute aspect.

4.1. Introduction

In the Rb atomic clock, the Physics package has a very decisive role. The Physics package comprises of Rb bulb, absorption cell, and photo-detector, these are all temperature sensitive and each requires a specific temperature for the optimal operation. The optimal temperature for the three devices are not the same, so the temperature is a compromise of the desired temperatures, making the devices more sensitive to the external temperature changes. The transition frequency of the Rb atoms is sensitive to the pressure in the gas cell, which changes with temperature. For proper operation, the absorption cell and lamp have the specific temperatures for a zero temperature coefficient condition. The deviation from these temperatures means the Rb clock becomes more susceptible to the ambient temperature changes. The important requirement of the Rb clock is to maintain the Rb bulb and absorption cell at the operating temperatures 110°C and 75°C respectively, even for the scenario of the surrounding temperature varying from 0°C to 15°C in thermo-vacuum condition. The thermal design of the Rb clock is aimed to provide the required temperature stability for the Rb bulb and Rb absorption cell assemblies under different operating conditions. This is achieved using mainly, the active thermal PID control techniques augmented with the electrical passive heaters.

4.2. Thermal control systems

The thermal design of the Rb clock is arrived at using the various passive thermal control elements like tapes, coating, spacers, multi-layer insulation blankets etc. It is augmented using active thermal control technique like auto-commanded heaters (PID controller). All the internal parts except magnetic shields, are provided with the low emissive tape. The microwave cavity is isolated from photodiode detector PCB holder and hence from the main housing, using thermal isolation washer support brackets. The identified thermal contacts provide required thermal conductance, sufficient to maintain the internal temperatures. The Physics package of the space Rb clock has a base plate, which is isolated thermally from the deck using appropriate thermal isolation washers (Titanium washers). The interface temperature at the Physics package base plate and S/C deck is to be maintained between 0°C and +15°C with ±1°C accuracy. The temperature stability requirement for the Rb bulb assembly and absorption cell assembly is of the order of few milli °C. The temperature controls are provided at four places. The Rb bulb housing and absorption cell housing are maintained at 110°C and 75°C respectively. With the help of two additional temperature controllers, temperatures at 70°C for Main housing (for Rb bulb and cavity) and 30°C for base plate respectively are maintained with the temperature stability of the order of few milli °C.

4.3. Overall power budget of physics package

The power required to stabilize the Physics package temperature is 36 W at warm up and 16 W at steady state in the final design. The lamp oscillator circuit dissipates 4.5 W and its heater requires 1.5 W for maintaining it at 110°C. The control circuit is based on a

Table 4.1 Estimated power budget for Rb atomic clock.

Sr.No.	Description	Maximum Power (W)
1	Rb Bulb — Dissipation	4.5
2	Rb Bulb — Heater	1.5
2	Absorption cell — Heater	3.5
3	Main Cylinder — Heater	10
4	Base Plate — Heater	14
5	PID temperature controller circuit	2.5
	Total	36

small PID temperature controller running in the analog mode. This controller is able to run all the temperature control circuits with a power dissipation of 2.5 W. These controllers keep 10 W for the common cylinder and 14 W for the base plate. The microwave cavity needs 3.5 W to maintain its temperature at 75°C. A summary of the final clock power budget is outlined in Table 4.1.

4.4. Thermal simulations and analysis

Thermal Mathematical Model (TMM) based final layout of the Physics package is developed and analysed. After developing the geometrical model, the appropriate boundary conditions are simulated to evolve TMM in I-DEAS TMG software. The software ANSYS is used for performing the finite-element simulations for the Physics package and monitoring power requirement, for setting the specific temperature for each subsystem. The simulations are performed on the module hypothesizing different configurations. The configuration for the module equipped with Rb bulb and the configuration for the microwave cavity and the external common cylinder are simulated. Finally, the entire Physics package configuration with three layer magnetic shields is simulated. The temperature distribution along the heated area is observed in the different cases. All these simulations are performed hypothesizing no losses other than conduction in the cavity walls (complete vacuum, no radiation). Finally, the effect of convection on the heating and the losses that it introduces is also estimated, simulated and experimentally validated. The final configuration simulated is the module in its original package with the cavity and lamp assembly mounted on common cylinder with three magnetic shielding layers, that introduce a considerable conductivity loss. The outside environment zone ("cold" zone) is hypothesized at 25°C. In this thermal design, the spacecraft deck is the only common heat-sink available. The temperature of all the critical parts in the Physics package is affected by the deck temperature. Hence the cold and the hot aspects of deck are considered for extreme conditions. The deck temperatures $T_{deck} = 0°C$ and $T_{deck} = 15°C$ are considered as extreme conditions. The thermal performance for these extreme conditions is discussed here. Overall TMM developed is shown in Fig. 4.1 with MLI and without MLI and side covers. A fine mesh is used in this case, in order to carefully observe what happens at the cavity edge. TMM details of Rb bulb assembly and RF microwave assembly are shown as a screenshot of the meshed device in Fig. 4.2. In the simulation, it is possible to clearly see the structures which are finely meshed in order to estimate the temperature distribution along these areas. The internal heat dissipation in Rb bulb assembly is taken to be total 4 watts and there is no internal heat dissipation in RF microwave assembly.

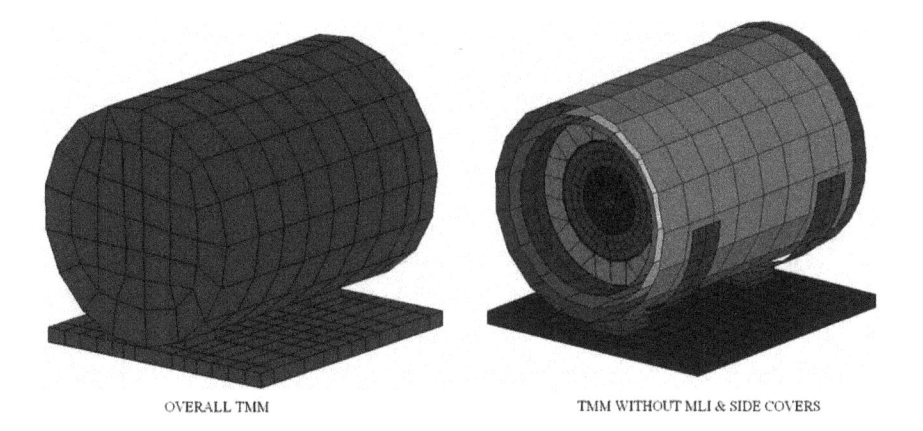

Fig. 4.1 Assembled view of developed thermal mathematical model.

Rb bulb assembly **RF Cavity Assembly**

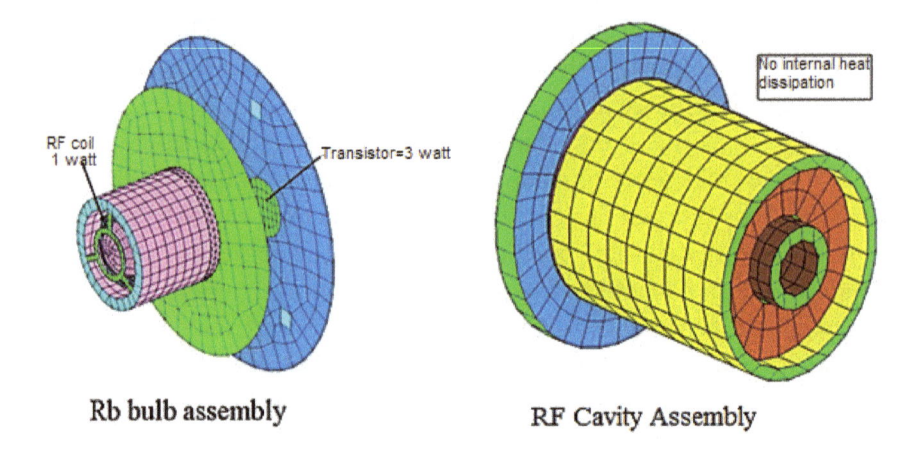

Fig. 4.2 TMM details of main assembly.

4.5. Heat dissipation problem in Rb bulb

In addition to the power dissipated to heat the Rb absorption cell, the Physics package has one more important component that dissipates power. The lamp's DC power supply dissipation is about 6 W, and the RF power required to excite the bulb at 81MHz is ~1.5 W. The lamp power consumption is unique in the sense (a) the most of the power must be dissipated as heat by the structure supporting the lamp and (b) the lamp must be operated at a constant temperature in order to maintain stable light intensity inside the absorption cell, and reduce time-varying AC Stark shifts of the clock output frequency. We maintain the lamp at a constant temperature by using a small heater directly under the lamp housing, where lamp is fixed with sil fill epoxy shown in Fig. 4.3. However, the lamp is heated by its own power dissipation even with no power input from the heater. We assume that the ambient temperature can be in a range of T_{min} to T_{max}, and that the bulb is kept at a constant temperature $T_L > T_{max}$. If the thermal conductance between the lamp and the environment is constant and it is assumed that no heater current is used to heat the lamp when $T_{ambient} = T_{max}$. Then the power required to heat the lamp for $T_{ambient} = T_{min}$ is given by

$$P_{max} = \frac{(T_L - T_{min})}{(T_L - T_{min})} P_{min}, \quad (4.1)$$

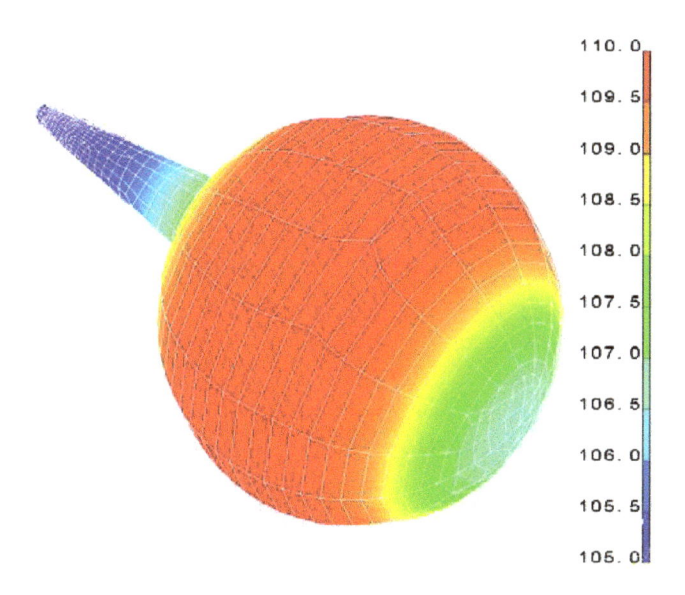

Fig. 4.3   Temperature distribution on the Rb bulb.

where P_{min} is the power dissipated by the lamp alone. This implies that if $P_{min} = 6\,W$, $T_{min} = 25°C$, $T_{max} = 30°C$ and $T_L = 110°C$, then the maximum power dissipation $P_{max} = 6.375\,W$. Because of the large power dissipated at the minimum ambient temperature, it is clear that this design is not ideal. The lamp should probably be in thermal contact with the microwave cavity, in order to use the power dissipated by the lamp to heat the entire Physics package. Equation (4.1) describes the maximum power required to heat the Physics package and not just the lamp. There are several other approaches that can help address this difficulty. One is to increase the operating temperature of the lamp while simultaneously decreasing the thermal conductance between the lamp package and the environment. The second is to reduce the power dissipated by the lamp by running it close to threshold.

The electrical power required to run the Physics package, including the base-plate heating, is 36 W and is dominated by the power required to heat the base plate. A total of 3.5 W is required to heat the absorption cell to 75°C, which is 45°C above the base plate temperature of 30°C. The base plate temperature is maintained above ambient in order to give a reference for the Physics package. By modelling the heat flow in the structure, both analytically and with a finite-element computation, the heat loss channels could be broadly identified. We estimate the loss through the three aluminium spacers, each with a diameter of 100 mm and length of 60 mm. The spacers provide the electrical connections between magnetic shields to the base plate. The remainder heat is presumably lost through radiation, and through conduction and convection in the air surrounding the Physics package.

4.5.1. Thermal simulation of Rb bulb

In RF driver for Rb lamp and its ballast, how to improve the efficiency and cool the heat dissipative components, are the formidable problems for the design engineer. These Rb lamps bring many challenges to the design of the integral systems. For example, consider a typical application, where the housing is totally enclosed, and the lamp heats the environment of the ballast by conduction through the copper housing and the radiation. Figure 3.18 shows all the parts of the Rb bulb assembly with the RF oscillator circuit. The RF exciter coil is wound around the Rb bulb. The Rb bulb sits on a Teflon bush surrounded by a cylindrical copper housing. The gap between exciter coil and bulb housing is filled with high thermal conductive epoxy (Sil Filler 2000). The Rb bulb housing along with the Rb bulb, RF exciter coil, bush and epoxy, are soldered to PCB at 3 copper pins. The PCB along with the Rb bulb housing assembly is fixed on PCB mount, which is made of aluminium. A heater with temperature sensors is mounted, and an appropriate thermal design is worked out to meet the temperature requirement. The finite-element simulation shows an inhomogeneous distribution of the temperature. The Rb bulb is maintained to nearly 110°C, the functional requirement of the bulb. The temperature distribution on the Rb bulb is shown in Fig. 4.3.

It is clear from the Fig. 4.3, that bulb temperature is the maximum at the centre and the minimum is at the tail. For a temperature of 110°C, typical working temperature of the Rb bulb, a 6 W power supply with 20 V and 300 mA is necessary. The next step is to simulate the bulb assembly device in its low-loss configuration.In this configuration, the central bulb assembly is modified and the module is equipped only with the 4 layer PCB, mandatory for letting the heat flow from the mounted transistor through the bulb holder. The module in this configuration is more fragile, but it can still withstand, without problems, the weight of a copper coil and the bulb assembly. On the other hand, it offers better performance in terms of heating. This module allows the alteration of the temperature distribution along the heated area, through the copper coil, that regulates the temperature flow on the bottom of the bulb notch. It is desirable to have a colder zone at bulb notch for keeping the alkali metal away from the radiating area. Another finite element simulation is performed hypothesizing that the value of the coil resistor, that regulates the flow of heat into the bottom part of the PCB ground, is equal to nearly 0 Ω. This means that there is no current flow on the upper part of the bulb, so the temperature gradient along the heated area is the maximum. The result of this simulation is shown in the Fig. 4.3. The temperature difference achieved between the bottom and the top of the heated area is in in the range of 5°C This happens because the bulb housing introduces a considerable conduction loss between the heated zone and the environment, hypothesized at 25°C. The oscillator PCB temperature is reached up to 105.3°C. The temperature of all other parts of this assembly is in the acceptable limits. Figure 4.4 shows the temperature distribution of the Rb bulb assembly with PCB.

4.5.2. Thermal simulation of absorption cell and microwave cavity

The various thermal packaging structures are studied in order to reduce the temperature sensitivity of the Rb absorption cell, with respect to change in ambient temperature. This sensitivity reduction is aimed to stabilize the cell temperature and to lower the power consumption. In this framework, the thermal management of the Rb-vapour cell takes into account three heat transfer processes,

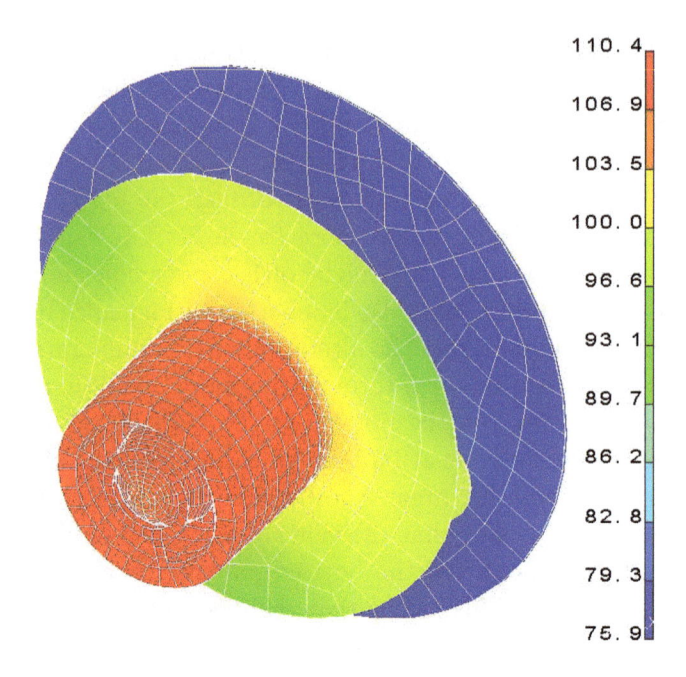

94 Rubidium Atomic Clock: The Workhorse of Satellite Navigation

Fig. 4.4 Temperature distribution on Rb bulb assembly.

namely, conduction through physical parts, convection between central cavity temperatures controlled part and magnetic shielding enclosure, and radiation from the heated absorption cell. To reduce the losses through these three processes, the gap between the absorption cell and the microwave cavity is filled with the soft thermal conductive epoxy material (Sil Filler 2000). The microwave cavity is mounted, on photodiode detector PCB holder, and is thermally isolated from the main housing, using thermal isolation washer/Teflon ring. This Teflon ring creates a high resistive path between the cell and it's surrounding, thus reducing the heat loss by means of conduction. Then, to overcome heat loss due to convection process, the complete Physics package is enclosed in vacuum. Finally, silver coating deposited onto the cavity inner surface surrounding the cell decreases radiation. Considering these solutions, thermal analysis is performed by finite element method and it is observed that the maximum temperature variation across the absorption cell is 1 to 2°C shown in Fig. 4.5. Thus, the power consumption to maintain the

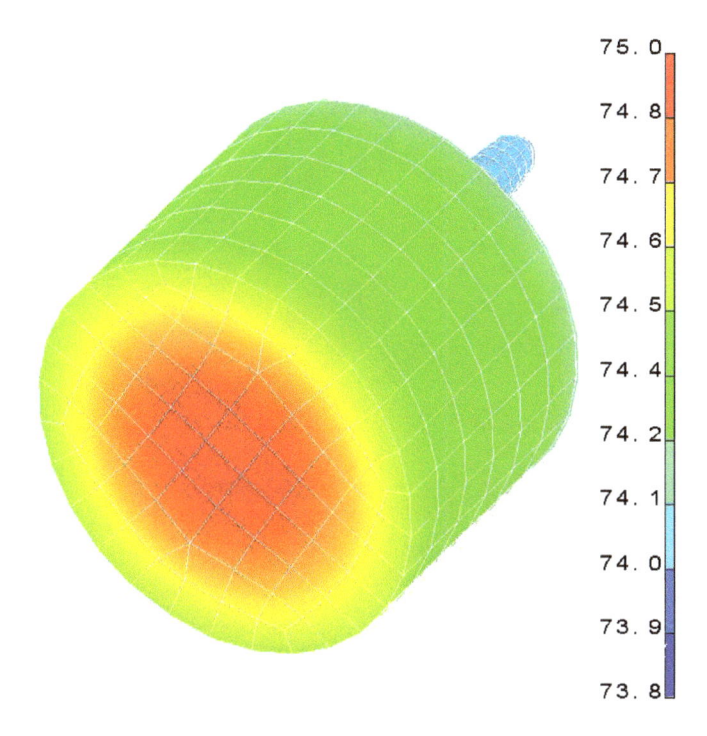

Fig. 4.5 Temperature distribution (10°C) on absorption cell.

cell-temperature at its operating temperature 75°C is of the order of 3.5 W for the ambient temperature 0°C.

It is possible to see that the temperature distribution is more homogeneous, and only the zones close to the Rb lamp are hotter than the other parts. The maximum temperature is on the Rb bulb facing side of the absorption cell. The tail of the cell shows nearly 1°C less temperature than the cell. Initially, a total power of 3.5 W (20 V and 175 mA) is necessary to achieve 75°C, with the aluminium washer between the cavity and the common cylinder. The conduction loss is decreased due to the addition of the 4 intermediate teflon bridges between the cavity and the common cylinder, which create a thermal open-circuit between the heated area and the cold zone. We get more or less the same temperature gradient as in the first case, when the absorption cell inside the cavity is in its original configuration but for achieving the same temperature, we need around 25% less power. The

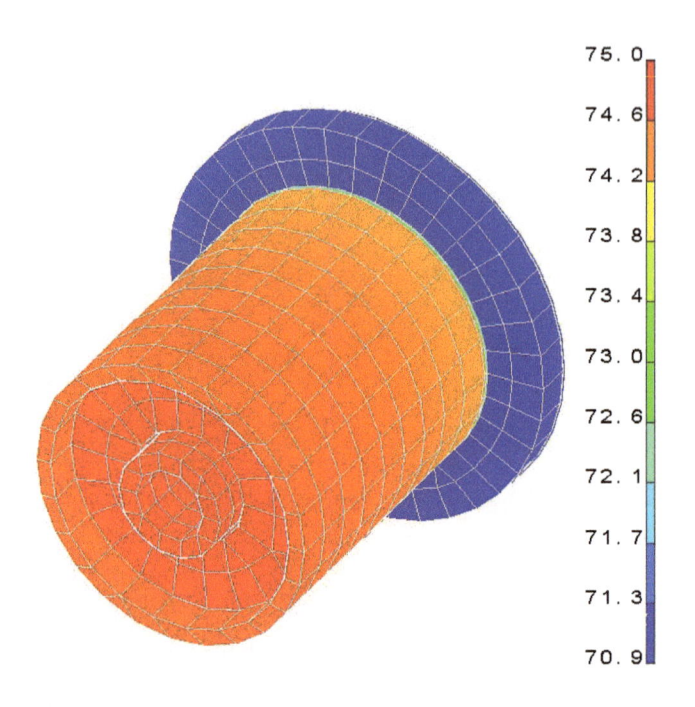

96 Rubidium Atomic Clock: The Workhorse of Satellite Navigation

Fig. 4.6 Temperature distribution (°C) on absorption cell assembly.

temperature distribution on the absorption cell with cavity assembly is shown in Fig. 4.6.

4.5.3. Thermal simulation of common cylinder and base plate

The main housing consisting of the Rb bulb and the microwave cavity has temperature ranging from 69.4°C to 72.5°C. The temperature distribution on the main housing is as shown in Fig. 4.7. The higher temperature is near the flange, where the Rb bulb assembly is mounted. In this assembly, a total of 4.5 watt power is dissipated and hence the main housing near this flange has the maximum temperature.

The base plate is maintained at nearly 30°C. The temperature distribution on the base plate is shown in Fig. 4.8. The minimum temperature on base plate is near the interface of the base plate with the spacecraft (S/C) deck.

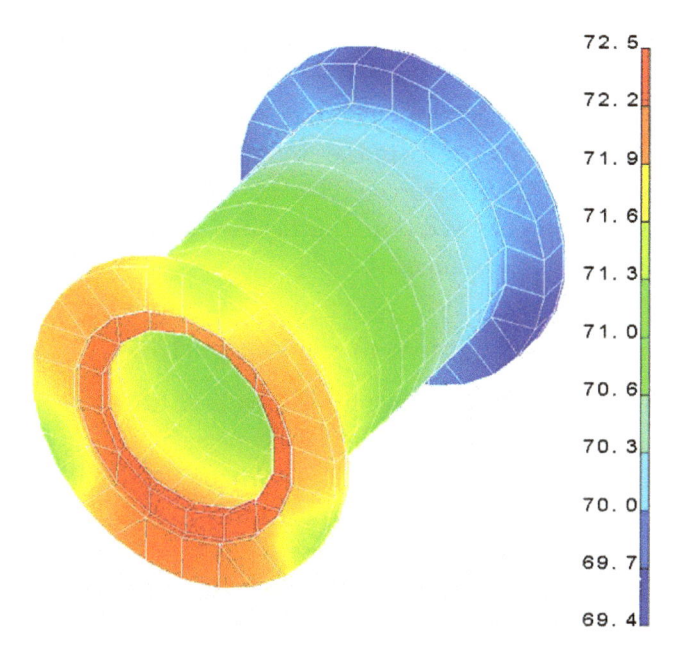

Thermal Analysis of Rb Physics Package Components 97

Fig. 4.7 Temperature distribution on main housing.

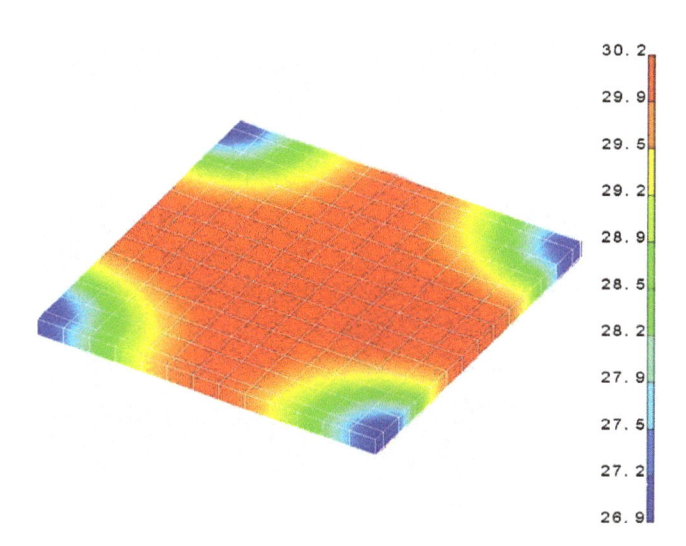

Fig. 4.8 Temperature distribution on base plate.

Based on the thermal analysis carried out on the Physics package with this thermal design, the functional temperature requirements are met for both the sub-assemblies i.e., the Rb bulb and the absorption cell. The heater capacity in the steady state condition is decided, based on the feedback from the Thermo-vacuum test. It should be such that from a cold start to the desired temperature, the time required is less that 50 minutes.

4.6. Heat loss mechanism in physics package assembly

The main sources of heat loss are conduction from the absorption cell to the base plate through the lower spacer unit, the conduction and negligible convection to the air surrounding the Physics package and the conduction in the suspension tethers. The heat loss to the frame and the package occur through radiation. The conduction to the air surrounding the package could be reduced to a negligible level by packaging the Physics package in a vacuum enclosure. However, even at atmospheric pressure, convection is suppressed due to the small size of the gap between the heated device and the cavity package. If the gas pressure in the evacuated enclosure is significantly below ~ 1 Pa (10^{-3} Torr), the thermal conductivity should be negligible. The radiation is dependent on the surface area and the emissivity. The gap dimensions, gas composition and the pressure drives the gas conduction. The heat loss due to the conduction is characterized, to determine how the vacuum level affects required heater power. For low pressures (less than about 20 mTorr) radiation is the dominant mode of heat dissipation. At high pressure, the heat transfer essentially depends on the thermal conductivity of the gas. The expected power dissipation is based on the Eq. (4.2).

$$Q = I(T_1 - T_0)\frac{A}{L}. \qquad (4.2)$$

Here I is the thermal conductivity of the material, A is the material's cross-sectional area and L is the material length along the direction of heat flow.

4.6.1. Assembly of sensors and heaters

The absorption cell temperature is maintained through an integrated single-element resistive platinum temperature-sensor and heating elements. The temperature sensor is distributed uniformly across the cavity face to provide an accurate average temperature measurement of the vapour cell. The power is delivered efficiently from the control electronics, while still maintaining sufficient voltage overhead to provide fast device turn on and good control response for different sub-systems. Both the heating and the sensing resistors for cavity are configured in such a way that the current flowing through one segment of the resistor is balanced by the opposite direction current in another segment, thus minimizing the magnetic fields in the vicinity of the vapour cell. The significant improvements in the thermal engineering reduces the power requirement to nearly 3.5 W.

4.6.2. The radiation heat loss in the Rb lamp assembly

Detailed models of the radiation must account for the multiple surface emissivity and shape factors of the system. However, the heated Rb bulb has radiative heat loss with a temperature dependence given by Eq. (4.3),

$$P = \varepsilon \cdot \sigma \cdot A_{\text{Cell}} \cdot (T_{\text{cell}}^4 - T_{\text{amb}}^4) \quad (4.3)$$

Where ε is a parameter which embodies the relative emissivity of the cell and heat-reflecting package, A_{cell} is the surface area of the cell, and σ is the Stefan–Boltzmann constant. As a general rule, the heat loss by thermal radiation is reduced by minimizing the bulb's surface area and average emissivity. The surface of the bulb is partially covered with Teflon and copper. These materials have emissivity around 0.5 with variation depending on the surface finish. Some materials such as polished copper can have emissivity as low as 0.03. If a gold coating, with emissivity 0.02, is placed on the interior surface of the shield, the heat transfer from the Rb bulb to the common cylinder, the absorption cell and PCB, due to radiation may be 13 mW, 6 mW and 16 mW respectively, for a temperature difference of 25°C.

4.7. Experimental setup for thermo vacuum test

Four temperature control circuits are used for controlling the temperature of the entire Physics package. The first temperature-controlled element is the aluminium cylinder that holds the lamp oscillator and cavity assembly. The cylinder is wrapped in a foil heater with a working temperature a 65°C. The cylinder is placed, inside three concentric layers of mu-metal magnetic shield. The absorption cell is thermally coupled to the cavity, whereas the tip of the stem containing the Rb metal is in contact with the external thermal shield. The details are given in Chapter 10. The temperature stability of the cavity and the external thermal shield, is of the order of 1 mK, for the integration time up to 1 day. The established temperature gradient of 2°C, between the stem and the cell body is required to limit condensation of Rb on the inner surface of the cell. The geometry of the cell is very important. The stem, in particular, plays a key role. It is approximately 2 cm long and has an internal diameter of 3 mm. The Physics package is placed in a vacuum chamber to isolate it from environmental perturbations. The entire assembly of the Physics package is shown in Fig. 4.9.

Fig. 4.9 Assembly of physics package for thermo vacuum test.

4.8. Thermo-vacuum test cycle

This section gives the details of the Thermo-vacuum test of the design verification model of the Physics package. Thermo-vacuum test is carried out as per the Thermo-vacuum test plan given in Fig. 4.10. Here, the steady state temperatures are attained at all the four critical locations for every interface plate temperature of −5°C, 0°C, 5°C, 10°C and 15°C. In each case, warm-up period of different subassemblies and the temperature variation at the steady state condition are measured. Test results for interface plate (IP) temperature of 5°C are considered. Once the readings for IP = 0°C cycle is over, heaters and power supply to RF circuit is turned "OFF" and IP temperature is raised to 5°C. It is achieved after 55 minutes. Then after about 17 minutes, the heaters and the power supply to RF exciter circuit are turned "ON". The time taken to reach the required temperature for four controlled parts, is shown in Table 4.2.

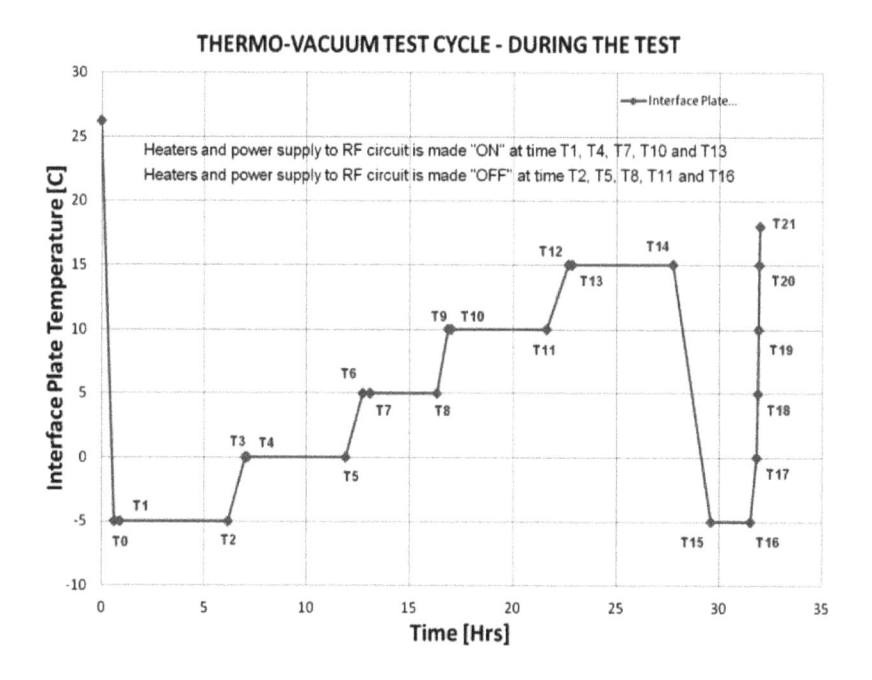

Fig. 4.10 Thermo-vacuum test cycle during the TV test.

102 Rubidium Atomic Clock: The Workhorse of Satellite Navigation

Table 4.2 Warm-up period of different sub-assemblies for IP temperature = 5°C.

Sr.No.	Component	T Initial	T Final	Time (Final)	Duration, HH:MM
1	Interface Plate	4.9	4.9	—	—
2	Rb Bulb	41.9	153	00:40	01:18
3	Absorption Cell	42.8	74	00:36	01:14
4	Main Housing	39.7	67	00:07	00:45
5	Base Plate	21.8	37.1	01:35	02:13

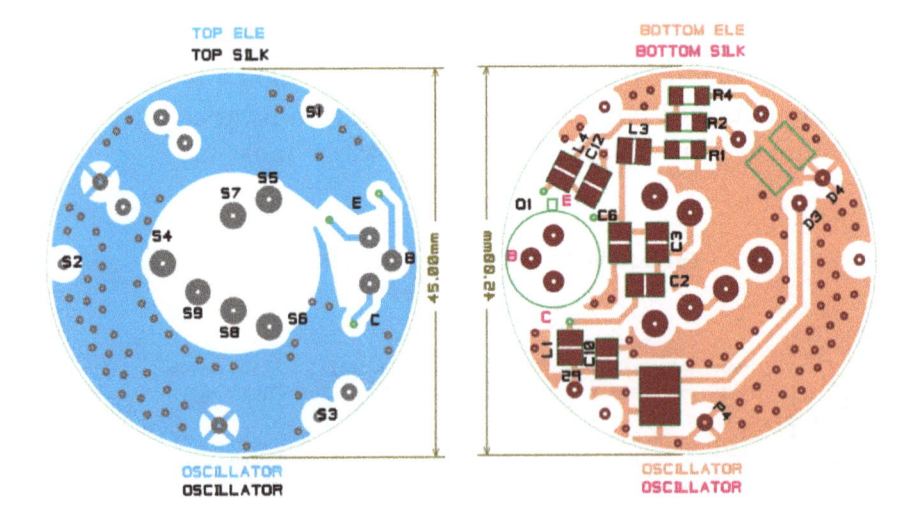

Fig. 4.11 Layout of DVM lamp driver PCB.

It can be seen from the above Thermo-vacuum results, except the bulb assembly, that the required temperature is attained and maintained very well in all the other sections, with the good stability. The bulb assembly has a higher than the expected temperature. Hence, after the test the Rb bulb assembly of DVM model is inspected. The top and bottom copper layer's thickness of the actual PCB, is 70μ each. The PCB is mounted on the PCB support plate at locations S1, S2 and S3 and it is clearly visible that the copper content is not available on the bottom face of the PCB, as shown in Fig. 4.11.

Thermal Analysis of Rb Physics Package Components   103

Table 4.3   Power dissipation in lamp assembly.

Sr.No.	Component	Heat Dissipation, Watts	Measured Power
1	RF coil	0.45	Voltage (V) = 18.5 Volts
2	Transistor	2.89	Current (I) = 210 mAmps
3	Resistor	0.56	Power Dissipation = V × I
	Total	3.9	Power Dissipation = 3.9 watts

The thermal analysis is carried out with these PCBs, incorporating the actual boundary conditions during the Thermo-vacuum test. The heat dissipation values in this test are given in Table 4.3. These values are also incorporated in the thermal model.

Based on the thermal analysis, the PCB layout design is modified to meet the thermal requirement. We add two extra bottom copper layers (thermal layer) with 70μ thickness and make a 4-layer PCB. The extra copper is added on top layer of PCB, where Teflon bush is fixed. Thermal layers are now in contact with the top and bottom copper layers. The copper area in bottom layer is increased to the maximum extent and is in contact with the mounting lugs for PCB support. The PCB mounting lugs are plated through hole (PTH) type and in contact with the top, bottom copper and internal thermal layers. The TMM is updated with the new PCB design. After the modification, the entire clock is tested in Thermo-vacuum condition. The thermal performance of each sub-assembly is also recorded. The temperature profile for different temperature controlling sub-assemblies is shown in Fig. 4.12.

The estimated heater power for the thermal control, during this test is based on the measured value of voltage and current at a given time for different IP temperatures. There are two sets of power supplies, one is for the Rb bulb assembly and RF microwave assembly heaters, and the other is for the main cylinder and base plate heaters. Table 4.4 gives the estimated heater power used to control the temperatures during Thermo-vacuum test.

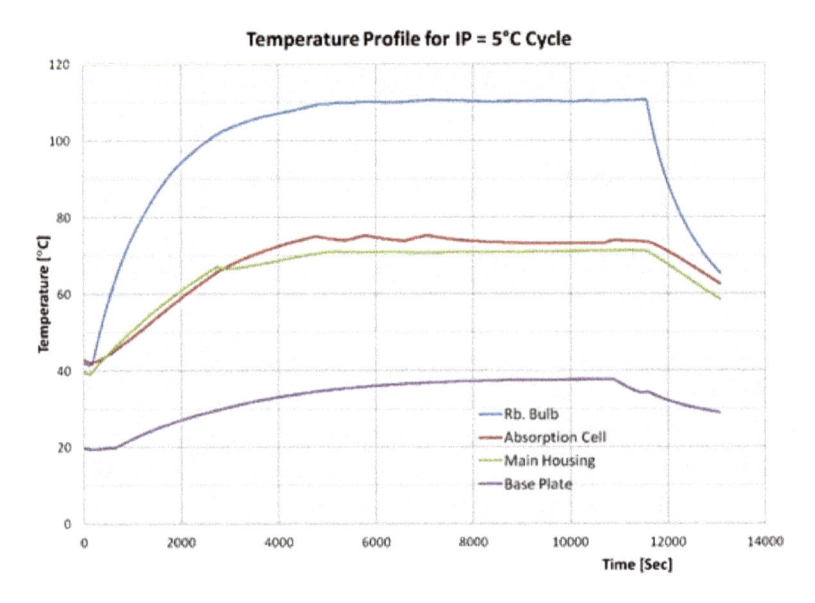

104   Rubidium Atomic Clock: The Workhorse of Satellite Navigation

Fig. 4.12 Temperature profile for various sub-assemblies during IP = 5°C cycle.

Table 4.4 Heater power estimation during the Thermo-vacuum test.

IP Temp	Set 1 of Power Supply			Set 2 of Power Supply			Total Heater Power, Watts
	V, Volts	I, Amps	Q1, Watts	V, Volts	I, Amps	Q2, Watts	Q1 + Q2
−4.8	23	0.125	2.875	28	0.4	11.2	14.075
0.9	23	0.175	4.025	29	0.4	11.6	15.625
5.5	23	0.175	4.025	29	0.4	11.6	15.625
9.9	23	0.175	4.025	29	0.35	10.15	14.175
15.1	23	0.125	2.875	29	0.3	8.7	11.575
−5.1	23	0.175	4.025	29	0.425	12.325	16.35

4.9. Results

It is concluded from this test that the thermal design, worked out with 4 Watt heat dissipation, is sufficient to control the temperatures of the critical components. The assumed values of thermal resistances are found reasonable. The heater capacity is to be kept as per the thermal design so that the required temperatures of base plate and main cylinder can be attained at lower IP temperature. The performance of the PID temperature controller for various sub-assemblies works as expected.

Chapter 5

RF Synthesizer for Rb Atomic Clock

> The RF synthesizer should have low noise multiplier chain and with suppression of spurious signals. Several schemes are used by different groups. The optimized scheme is discussed here.

5.1. Introduction

Rb atomic clocks have the potential for achieving frequency stability and accuracy better than 1×10^{-14} and 1×10^{-12} [86–92] respectively. One of the difficulties in realizing this performance in the passive atomic standards, is the development of a good frequency stable resonant microwave oscillator, for interrogating the field independent atomic g.s hyperfine transitions. Several of the clocks or frequency standards require an interrogation cycle, that ranges from 1s to 100s. Dick, *et al.* investigated in detail the frequency instability in a frequency standard that is caused by noise in the local oscillator [93]. In situations, where there are two or more atomic samples, it is possible to shape the interrogation cycle to significantly reduce the contribution of the local oscillator to the white frequency noise level [93].

A low phase noise, high resolution frequency synthesizer is needed for the atomic interrogation. Previously, several researchers worked on different topologies of frequency synbook for atomic interrogation. One traditional technique devised by R. Barillet [94], employs frequency multiplication with step recovery diodes (SRD). G.D Rovera [95] describes a topology realized by a frequency chain using sampling mixers to avoid narrow-band filtering, while R. Boudot [96] and F.R Martinez [97] utilize non-linear transmission line (NLTL) circuits instead of SRD for the frequency multiplication. However, the disadvantage of frequency multiplication is its large energy dissipation, leading to very low energy efficiency.

In this work, we describe the construction and evaluation of a low phase noise synthesizer for the 6.834 GHz microwave component of the clock interrogation. The phase noise measurements between two identical systems confirm no significant degradation of the 10 MHz local reference oscillator. The base frequency to the synthesizer is derived from a high stability voltage controlled crystal oscillator (VCXO), which guarantees the clock frequency stability better than $5 \times 10^{-12} \tau^{-1/2}$.

5.2. Synthesizer design methodology

In the case of Rb atomic clock, the frequency synthesizer generates 6.834 GHz clock interrogation signal by multiplication from a 10 MHz reference signal provided by (Fig. 5.1) quartz oscillator (Centum 10 MHz-SC Crystal Oscillator), which may operate as a stand-alone oscillator, having a relative frequency stability of the order of 3×10^{-12} for the sample time of 1s. The primary function of the internal 10 MHz oscillator is to provide good close-in spectral purity and time domain stability for time shorter than about 1s. The VCXO output with constant power, is distributed among the different arms of the synthesizer. The multiplication arm is shown in the Fig 5.1, where reference 10 MHz signal is 9×4 multiplied to 360 MHz using active transistor multiplier and band pass filter. Such active transistor based frequency multiplier presents several advantages with respect to other methods of frequency multiplication, such as low phase noise, wide range of input frequencies, low power, and high output power at high harmonics [98–99].

A separate section generates a 5.3125 MHz from FPGA based direct digital synthesizer. 5.3125 MHz signal is frequency tuneable and controllable in power, which is mixed to the 360 MHz signal and further multiplied by 19 times by SRD multiplier to constitute the interrogation signal. The amplification of the 360 MHz signal

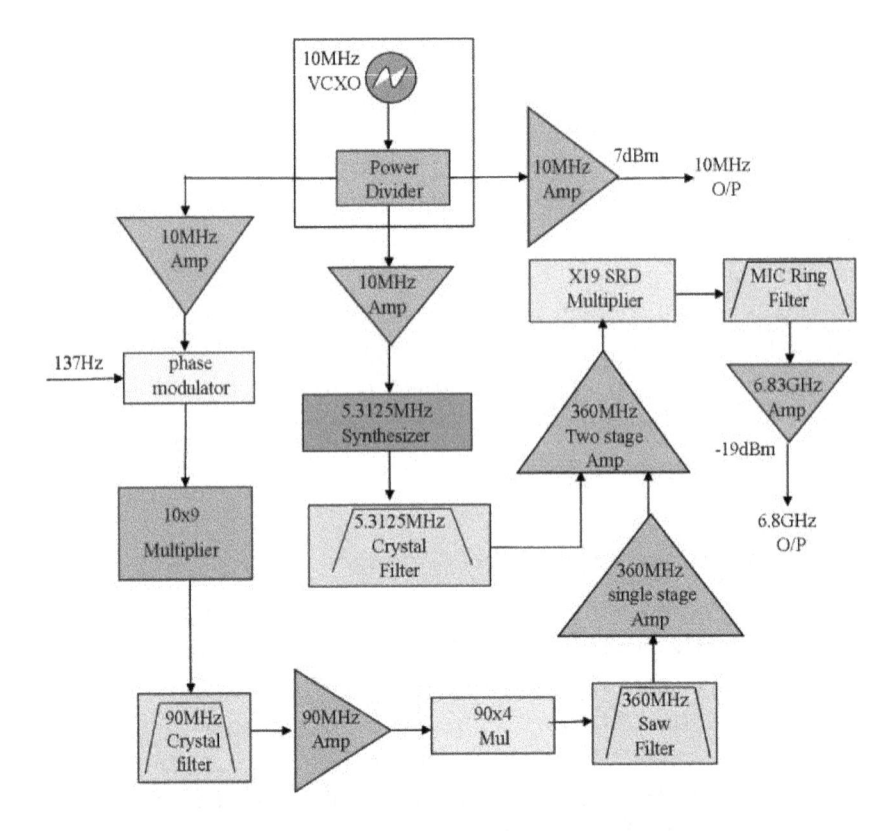

108   Rubidium Atomic Clock: The Workhorse of Satellite Navigation

Fig. 5.1   Block diagram of Rb frequency synthesizer.

provides the necessary power of 20 to 22 dBm to the SRD input. This amplification is probably the most sensitive section of the synthesizer. In the linear operation regime, the chosen amplifier's gain is slightly more than 21 dB. The 1 dB compression point occurs at an input value of 1 dB, where the amplifier output is approximately 22 dBm. The use of a higher input frequency and power to the SRD increases the power in each harmonic as well as in the harmonic spacing, which in turn simplifies harmonic selection and filtering. A band pass filter selects the desired harmonics and reduces the adjacent 360 MHz sidebands by approximately 40 dB. The sideband at 1 GHz are less than -70 dBc. The 19th harmonic is selected and amplified to provide -19 dBm at 6.834 GHz.

5.3. Precision oven controlled crystal oscillator

In the passive atomic frequency standards, such as the Cesium-beam device [100] and the optically pumped Rb-gas cell [101], the atomic transitions are probed with an interrogating signal derived from a crystal oscillator. We obtain a resonance signal, which is used for locking the frequency of the crystal oscillator to the transition frequency of the atomic ensemble (frequency-lock loop). In both Cesium and Rb clocks, one is concerned with the performance of the total system. In practice, this means that the locked quartz-crystal oscillator should have the best overall frequency stability. Another region of interest is the short-term region, where the limit of stability is normally set by the quartz-crystal oscillator itself.

In the present section, we verify the performance of a 10 MHz OCXO by comparing it with 10 MHz Cesium reference. The analysis is done by different types of quartz-crystal oscillators comparing to Cs atomic reference frequency. Based on the results, the good VCXO is selected as reference for the multiplier and synthesizer.

5.3.1. Phase noise, frequency sensitivity and short-term stability of OCXO

The OCXOs have very good short-term stability, but their aging affect the long term stability, in the free running condition. In atomic frequency standard, the stability is that of the OCXO, locked to atomic transition frequency via a servo loop. When locked, the frequency stability varies as the square root of the measurement time interval (τ) for the short integration time. The requirement of an atomic resonator is to have narrow line width (W) or high quality factor and high signal to noise ratio for good short term stability, defined in terms of Allan deviation,

$$\sigma_y(\tau) \approx \frac{0.2}{(SNR) \times Q} \times \tau^{1/2}, \tag{5.1}$$

where $Q = \dfrac{v_0}{W}$ and v_0 is atomic transition frequency. The Allan deviation $\sigma_y(\tau)$ defines the stability of the OCXO locked to the atomic transition frequency. It is also observed that the region, where $\sigma_y(\tau)$

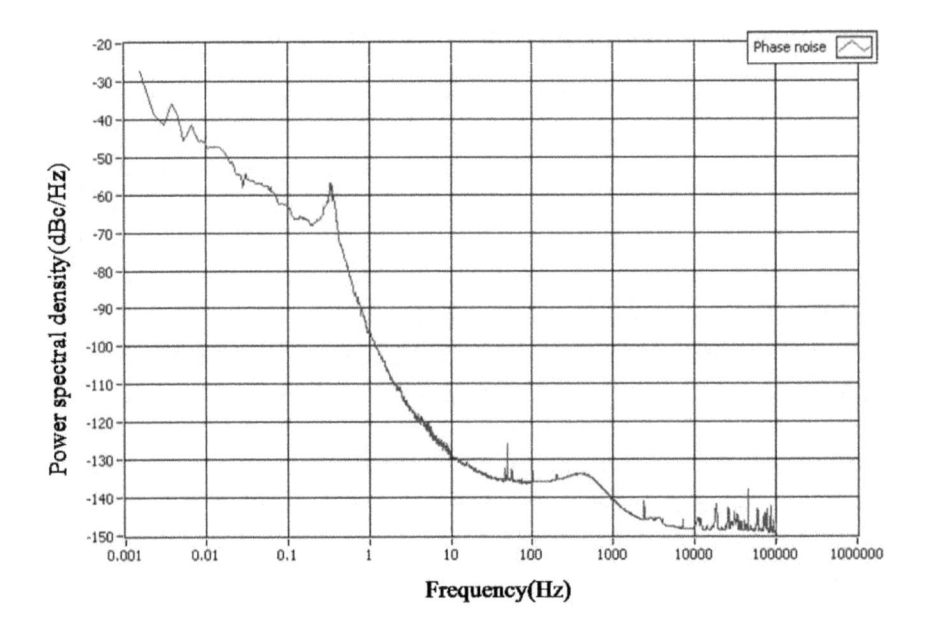

110 *Rubidium Atomic Clock: The Workhorse of Satellite Navigation*

Fig. 5.2 Phase noise plot for Centum OCXO locked with Cesium source.

is proportional to $\tau^{-1/2}$ occurs only when the loop bandwidth has a value of about 0.5 Hz. The optimization of the loop bandwidth is done and verified by phase noise plot of OCXO, locked with Cs reference, Fig. 5.2. The Best overall performance is also observed for the loop bandwidth of 0.5 Hz. For the larger values of loop bandwidth, the stability of the OCXO for $\tau < 1$ s is degraded by the white frequency noise of the atomic reference.

Finally, some small improvements in the stability of the locked OCXO in the short-term region are obtained at the expense of increasing the bandwidth of the servo loop. In the design of atomic frequency standards, it thus appears preferable to use a crystal oscillator with the lowest white phase noise. It is verified that the excess of phase noise in the PLL band-width is generated by noise coming from the on-chip sweep circuit. Indeed, lower phase noise is observed when the sweep circuit is switched off. Furthermore, measurements of the OCXO alone give a phase noise of -98 dBc/Hz @ 1 Hz offset from the carrier, which is much lower than the closed-loop noise at the same offset position Fig. 5.2. This shows that the limitation in the performance is due to the PLL system, especially from the sweep circuit noise instead that of the OCXO.

5.4. Design and simulation of 10 MHz × 9 active frequency multiplier

5.4.1. Introduction

Frequency multipliers are important building blocks for many applications, as well as for the measurement equipment. The inherent upper limit of the operating frequency in designing the amplifiers and the oscillators for microwave and millimetre-wave communication, wireless communications systems [103] and radar systems, leads to the frequency multipliers playing an important role in signal generation and frequency conversion. To build such highly portable, low power and low cost [104] systems, an efficient implementation of the local oscillator with frequency multiplier (LO) chain is required [104–108]. For the synthesizer of Rb frequency standard, the local oscillator frequency at 6.834 GHz is required to be generated with the 10 MHz reference source. The total multiplication factor of 684 is implemented in 9 × 4 × 19 sequence. For this particular application, achieving high efficiency with low DC power consumption is of prime importance and hence for multiplication of 10 MHz × 9 × 4, the active frequency multipliers are preferred to the passive multipliers [109].

The frequency multipliers can be broadly classified into the passive frequency multipliers employing diodes or the active frequency multipliers employing transistors. Despite excellent performance of the passive multipliers (step recovery diode), in terms of spectral purity and generating higher order multiplications (> 15) or prime multiplications factors [110–111], they suffer from high conversion loss, ×9 multiplier typically has a conversion loss of around 20 dB, and small output power. Thus, it requires both pre-amplification and post amplification (PA). Therefore, the major disadvantage of passive multipliers is the high drive powers (< 20 dBm) and the poor conversion efficiency. One aspect of the development of the active and high power frequency multipliers with broad bandwidths and high conversion gain is the need to combine the functionality of the passive multiplier with the output, thus, simplifying and improving the efficiency of the overall system. The widespread use of FET and BJT frequency multipliers in microwave engineering has its roots in at least three important facts. First, their ability to provide better isolation and frequency conversion gain, resulting in a multiplication efficiency exceeding the fundamental limit of $1/n^2$, where n is the order of multiplication [112–114].

Second, they require a relatively low input signal level (< 3 dBm) [112]. Third, they usually consume little dc power and dissipate little heat; this is an important advantage in space systems. The possible disadvantage of the active frequency multiplier for certain applications is that the non-linearity is not good for generating higher order harmonics. Typically, the active frequency multipliers are opted only for the applications, where multiplication factor is 4 or less, due to their higher efficiencies and low power requirements. For applications requiring multiplication orders of 8, 9, 10 & 12, successive active multiplication stages are preferred over passive multipliers [115]. The field effect transistors (FETs), high electron mobility transistors (HEMTs), step recovery diode, Schottky and varactor diodes are often employed in the design of frequency multipliers [116–119]. A number of papers have been published on the non-linear characteristics of the microwave active frequency multipliers [120–122]. At microwave frequencies; these techniques usually employ a non-linear device, generating the desired multiple of the fundamental input frequency. Gopinath *et al.* [123] utilized a detailed I_{ds} model for the FET, which describes the influence of various aspects of the non-linear device on its multiplying behaviour. Maas [124], on the other hand, applied a generic piecewise linear trans-conductance (PLT) model, where only I_{ds} clipping effects are considered, to develop a generalized multiplier biasing and drive criteria. O'ciardha *et al.* [125] combine the best of Gopinath *et al.* and Maas techniques. The goal is to study and, if necessary, to refine the multiplier design criteria of Maas, by applying the PLT model in a more generalized and more extensive harmonic analysis. However, another development [126] has demonstrated an alternative tripler-bias arrangement, which has the potential to perform with significantly better results than predicted by the earlier theoretical models. In many frequency multiplier design approaches, two main problems are addressed. A set of optimal bias/drive regimes of operation [123–126] is the first problem and the choice of the proper harmonic termination for improving of the frequency multiplier performance [127–131] is the second problem. The operating performance is improved by the proper selection of input and output circuit terminating impedances at the fundamental and the harmonic frequencies. Rauscher [127] performed a detailed study of the doubler behaviour based on the fundamental frequency drive level, the device output terminating impedance at the fundamental frequency, and the device input terminating impedance at the second harmonic frequency. Camargo [128] performed the most in-depth research in this area, concluded that the optimum MESFET doubler operation is obtained when the drain is terminated at the fundamental frequency by purely reactive circuit, which resonates with the transistor's output capacitance. Borg et al. [129] suggested that the optimum terminating impedance for the bipolar multiplier is a short circuit at the fundamental frequency. Thomas et al. [130] presented a quantitative optimization analysis of the active multiplier conversion gain and the spectral purity, governed by the fundamental and the harmonic terminating impedances and regions of non-linearity. Johnson et al. [131] presented the novel design and optimization techniques for the frequency triplers.

The work reported here have the following objectives:

(1) The simulated and experimental determination of the optimal bias/drive regimes (i.e. dc bias and input signal level) uses the modified non-linear model of a bipolar transistor which helps in terms of the known non-linear behaviour of transistor devices, for the design and also qualitatively explain the nature of the operating characteristics of the frequency multiplier by 9.
(2) Explain the techniques for the detailed study of harmonic termination of a bipolar transistor. The design and optimization is done on Agilent's Advanced Design System (ADS) with

harmonic balance (HB) as non-linear engine simulator. This combines the harmonic balance analysis with the full wave analysis for the planar circuits applicable for all simulations. It gives the detailed experimental measurements of the performance of a 10 MHz − 9 frequency multiplier, based on idler networks. The effect of the 9th harmonic input impedance on the conversion gain is simulated and experimentally verified. The design uses bipolar transistor BFY-193 as non-linear element along with the passive input and output circuit on the glass epoxy PCB.

5.4.2. *Design methodology of 10 MHz × 9 frequency multiplier*

For a multiplier depending upon the application, various design goals can be of importance, such as the maximum conversion gain, the maximum output power, high efficiency, high unwanted harmonic rejection and low VSWR. For this particular application, the fundamental frequency at 10 MHz is to be multiplied by 9 to generate the required output at 90 MHz. The different design goals are often conflicting in nature and a trade-off is often required. The primary steps involved in the design are (a) selecting the appropriate device and its biasing and (b) determining the optimal source and load impedance, and to provide the maximum P_{out} and conversion gain (CG), and synthesizing these impedances into low-loss input and output networks. The steps followed during the design are described in the following sub-sections.

5.4.2.1. *Devices selection*

The first step in the design process is to choose a suitable non-linear device. The non-linear properties of BJT, FET, MESFET, PHEMT or HBT can be utilized for generating the harmonics of fundamental frequency depending upon the input and output frequencies, noise and efficiency requirements. As the required frequency is less than 7 GHz, an unbalanced, single-transistor (BJT) topology using harmonic reflection and matching networks is chosen, as an active device for simplicity, low loss, cost and efficiency [130–131]. One key parameter that limits the performance of transistor frequency multipliers to relatively low input frequencies and multiplication orders, n, is the transistor's cut off frequency, f_T. As the current conversion efficiency for the output harmonic n is approximately $f_{T/n}\ f_o$ [114]. It is clear that one way to enhance the efficiency is to use a high f_T device. The high reliability is also an important criterion for selection, as the circuit is a part of the satellite payload. Thus, the device used as non-linear element in this multiplier design, is Siemens make silicon bipolar transistor BFY-193. The device is a space qualified microwave semiconductor, hermetically sealed in microwave package (Micro-X1) with f_T of 8 GHz. At 2 GHz, the device gives the maximum power gain of 13.5 dB. For the space based designs, reliability is of high importance and so the special care has to be taken in the design including de-rating.

5.4.2.2. *Bias optimization*

Although variety of non-linear mechanisms are present in the active devices, that can contribution to harmonic generation under large signal excitation, the primary source for generating harmonics is by clipping the collector current. This can be achieved by biasing the BJT base such that the collector current clips, when either the base voltage exceeds the forward conduction threshold or when the base voltage drops below the "turn-on" threshold voltage. This clipping can be achieved by appropriate selection of the device operating point. The criteria for deciding the operating point depends upon the required harmonic content at the output. The bias point is selected to enable the clipping of the input waveform either symmetrically or asymmetrically. When the output conducts for half cycle or less, the non-linearity is enhanced and output current contains train of harmonics. The conduction angle thus becomes the criteria for bias selection. The conduction angle is defined as the fraction of the fundamental input signal cycle (360°) for which the current flows into the collector of the device.

The base input signal level and the base bias voltage are selected to obtain a conduction angle that maximizes the output level of the desired harmonic. An approximation relating the harmonic collector current to the BJT conduction angle can be obtained using a Fourier series expansion for an ideal cosine pulse train [132] as,

$$I_n \approx I_{\max} \frac{4\theta}{\pi^2} \left| \frac{\cos n\theta}{1-(2n\theta/\pi)^2} \right| n \geq 1, \quad (5.2)$$

$$I_n \approx I_{\max} \frac{4\theta}{\pi^2} \left| \frac{\cos n\theta}{1-(2n\theta/\pi)^2} \right| n \geq 1, \quad (5.3)$$

$$I_{dc} \approx I_{\max} \frac{2\theta}{\pi^2}, \quad (5.4)$$

where I_n is the collector current for the nth harmonic, I_{dc} is the average value of the collector current, $I_{\max}$ is the maximum collector current, and 2θ is the conduction angle. The harmonic collector current relative to the maximum collector current from Eq. (5.2) is plotted in Fig. 5.3 as a function of conduction angle, for up to the

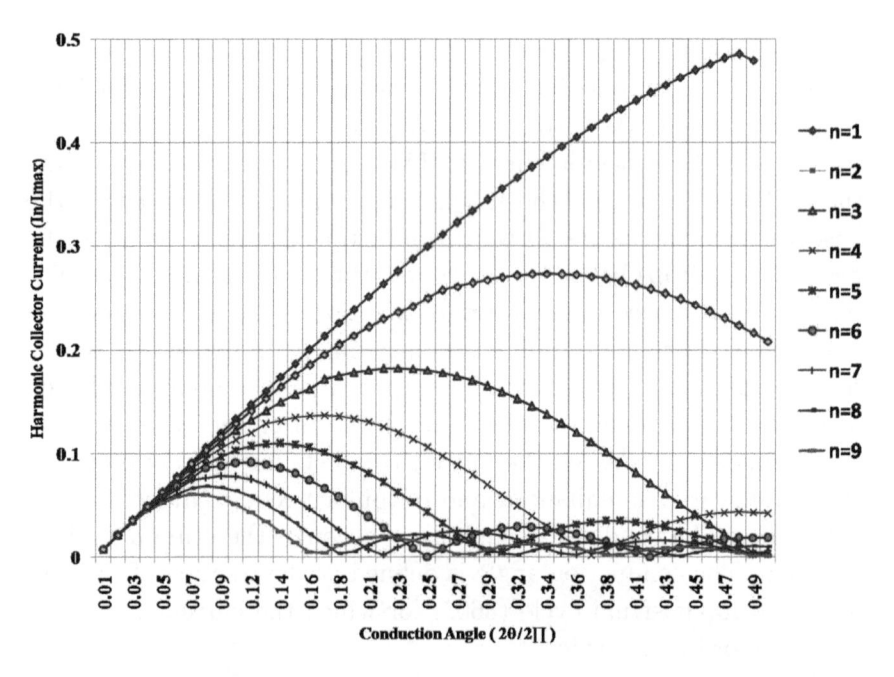

Fig. 5.3 Ideal harmonic collector current as a function of BJT collector conduction angle.

ninth harmonics. In order to maximize the desired harmonic output power, I_{max} should be as large as possible and θ should be chosen to give a high I_n. It is evident from the figure that, higher the harmonic required the shorter is the conduction angle that is to be maintained.

For the present case of $n = 9$, the optimal conduction angle is approximately $0.08 \times 360° = 28.8°$, which corresponds to $I_n/I_{max} = 0.06$. This clearly requires that the device to be operated in extreme cut-off region. For a circuit to be used in the space payload, certain de-rating guidelines are to be followed to improve the circuit reliability. Accordingly, the device cannot be operated at the maximum current and voltage ratings. Thus, by careful examination, the optimum operating point of the device is found to be 3 V, 10 mA. The output waveform is in pulse form and conducts for less than half of the input duty cycle. The collector current obtained is shown in Fig. 5.4. The device is operated at a conduction angle of 170°, which is clearly not in match with the theoretical value of 28.8° as calculated from Fig. 5.3. The device is not operated in the extreme cut-off to keep the harmonic current levels within the de-rated values.

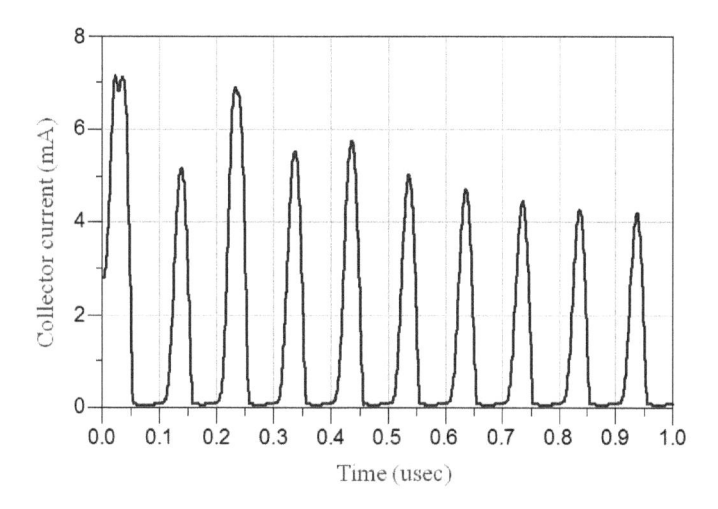

Fig. 5.4 Simulated collector current waveforms of 10 × 9 frequency multiplier.

5.4.2.3. Harmonic termination analysis

The frequency conversion occurs by making the transistors or diodes to operate in the strong non-linear region and generate harmonic signals. It is difficult to obtain relatively high output power at the desired harmonic frequencies. The filtering of unwanted harmonics may be done by simply using a band pass filter, but the harmonic rejection close to the carrier may not be sufficient. In the literature [116–118], there are varied opinions on how to optimally terminate a frequency multiplier. Nevertheless, there is a widespread belief that one of the best types of harmonic terminations is a short circuit. This can be obtained either by a filter network in the output or by the use of "idler" circuits in the output. The idler circuit provides a short at a particular frequency, hence the name — idler. The idlers are realized as short circuit resonators that are separate from matching and bias networks. This design makes use of series LC resonant traps called "idlers" for the desired harmonic, which allows termination of each harmonic in specific impedance without affecting other harmonics. Implementing the idler technique in higher order multipliers is quite challenging, since the number of L-C series resonant networks increase largely. The idlers are operated at unwanted harmonic frequencies causing the undesired harmonic currents to circulate in low-impedance reactive paths and thus, do not contribution to output power. By means of idlers (short circuit series LC resonators), both fundamental frequency and 9th harmonic impedances are optimized to achieve the maximum conversion gain. Therefore, the operating performance is improved by the proper selection of input and output circuit terminating impedances at the fundamental and the harmonic frequencies.

5.4.2.4. Syn-book of input source network

In the input side, the fundamental frequency signal is applied to the non-linear device through the matching network. The input network is also designed to provide a DC feed point to BJT, and DC blocking for base bias. In addition, an inductor and capacitor (LC) series resonator (idler) is inserted to properly adjust the 9th harmonic input impedance. The idlers are implemented as shunt resonator. The LC resonator not only reflects the 9th harmonic signal but can also be adjusted to maximize the desired output power, by taking into account its phase. Since the operating frequency of the resonator is defined by the multiplication of L_{res} and C_{res} values as shown in Eq. (5.5), the 9th harmonic impedance is easily optimized by changing these two values, while keeping the multiplication value constant to avoid any effect on the fundamental frequency impedance.

$$\varpi_0 = \frac{1}{\sqrt{L_{res}C_{res}}}. \tag{5.5}$$

For frequencies up to 2 GHz, carefully designed lumped element matching circuits can be used without any difficulty. The simple L-C resonators also have high Q to provide adequate rejection of unwanted harmonics.

5.4.2.5. *Syn-book of output load network*

The design of output network is done to pass the desired 9th harmonic, while suppressing the fundamental frequency and all unwanted harmonics. The output circuit provides a DC feed point to BJT and DC blocking for the collector bias and idler network. After careful study and examination of the harmonic contents of the device output, only specific frequency idlers are considered. The output circuit contains the idlers of 1st, 8th & 10th harmonics. The power of the close-in harmonics is specifically brought down by the use of two idlers at 8th and 10th harmonics. Additionally, as the power of the 1st harmonics is high, it has been shorted out using appropriate idler circuits.

5.4.3. *Multiplier circuits simulation design and development*

The design procedure of frequency multipliers involves controlling of the collector current harmonics effectively. Although this objective seems easy, practically it is quite tedious and tricky to implement.

The computer aided design (CAD) optimization is a very powerful tool to visualize and understand the non-linear effects prevalent in the multipliers. The harmonic balance (HB) simulation is quite effective and accurate in the analysis of strongly non-linear circuits. The accuracy of the approximation of the steady state non-linear behavior increases, as more harmonics of fundamental frequency are included in the analysis. However, it happens at the cost of higher memory requirements and the computational time. The HB analysis of higher order multipliers can be difficult due to the presence of strong non-linearities, large number of harmonics involved and possible instabilities, which make convergence difficult. The selection of the proper non-linear model for the device for the simulation purposes plays a very important role in designing a non-linear circuit accurately. The simulated design presented here is for a transistor ×9 frequency multiplier using the non-linear model of BFY-193 silicon bipolar transistor (Siemens make). The device is a space qualified microwave semiconductor, hermetically sealed in microwave package (Micro-X1) with Ft of 8 GHz. By the use of computerized design methods a high-accuracy Gummel–Poon model [133–135] of BFY-193 transistor is used for close performance predictions. The models and simulations presented here have been performed employing the Agilent Advanced Design System (ADS) software package. The design of the basic single-ended transistor based ×9 microwave frequency multiplier configuration, and the generalized circuit representation of the large-signal model are considered. The realization of an ideal single-ended ×9 frequency multiplier consists of an active device coupled with performance-enhancing input and output networks. The input network is designed to pass the fundamental frequency component and also to provide the required input matching. Similarly, the output network suppresses the fundamental and other undesired harmonics and at the same time providing the matching for the output frequency.

The first thing that is to be established is the stability of the circuit for a specific base and collector voltages. Then the input and output bias networks are designed by using high impedance LC filter at the input and output centre frequencies respectively.

The linear S-parameter simulation of ADS is used for this purpose. Although, the BJT frequency multipliers are practical for frequencies up to 4 GHz, they are often less stable. The BJT instability arises mainly due to two reasons (1) the low frequency gain of the BJT is much higher, which makes it prone to low frequency parasitic oscillations and (ii) under the strongly driven conditions, the reactive non-linearity of base to emitter capacitance may lead to strange parasitic oscillations. To avoid instability related issues, the series resistor must be carefully selected and the emitter must be properly grounded. The circuit is not stable over the full frequency range, so a resistor is connected in series to the base. The values of the resistors are adjusted until a stability factor greater than 1 is achieved. Care was taken to keep the value of R1 as small as possible so that it does not consume too much of the input power. For this simulation the ADS harmonic balance simulator and a non-linear model of BFY-193 are utilized. The BJT is operated near to the saturation as this bias is optimum for the 9th harmonic power. The peak reverse voltage across the base is also taken into account so that it did not cross the maximum specified limit of the BFY-193 used in the design. The 10 MHz to 90 MHz single-ended frequency multiplier without taking into account the adjustment of the 9th harmonic impedance is designed based upon the output impedance. Figure 5.5 shows the simulated output spectrum, for 0 dBm input power, with only bias network and decoupling capacitors. The fundamental harmonic content is quite high due to the device gain but the required harmonic level is very low (-16 dBm). For finding the optimal harmonic termination, a set of simulations of the short circuit combinations are performed. There is only one combination activated, the short circuit for the 9th harmonic at the input (simultaneously open circuit for the fundamental) and short circuit for the fundamental at the output (simultaneously open circuit for the 9th harmonic). Then an optimum output impedance at the 90 MHz, the 9th harmonic frequency for maximum output power from the BFY-193 is extracted using load-pull simulation. The resonator is placed in the input matching side and tuned. Three series LC resonant traps called "idlers" at 10 MHz (First harmonic), 80 MHz (8th harmonics) and 100 MHz

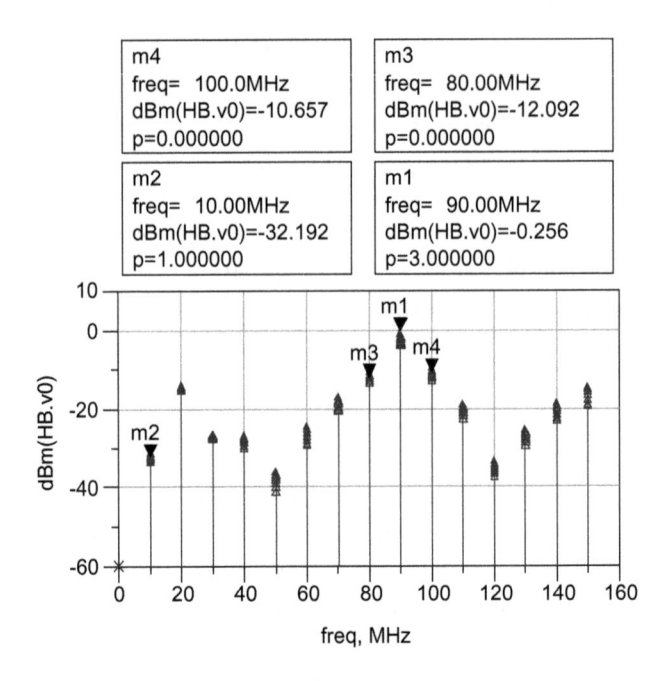

122 Rubidium Atomic Clock: The Workhorse of Satellite Navigation

| m4 freq= 100.0MHz dBm(HB.v0)=-10.657 p=0.000000 | m3 freq= 80.00MHz dBm(HB.v0)=-12.092 p=0.000000 |
| m2 freq= 10.00MHz dBm(HB.v0)=-32.192 p=1.000000 | m1 freq= 90.00MHz dBm(HB.v0)=-0.256 p=3.000000 |

Fig. 5.5   Simulated output spectrum without idler network.

(10th harmonics) are connected at the output to provide rejection of the unwanted harmonics in the output. A 10 MHz LC resonator line in the output network operates as a simple fundamental frequency reflector. For frequencies up to 2 GHz, carefully designed lumped element matching circuits can be used comfortably. The simple L-C resonators have high Q to provide adequate rejection of unwanted harmonics, as shown in Fig. 5.6. The topology of the idler is selected because of the available space for the circuit; any other topology may also serve the purpose if the size of the circuit is of major concern.

The large signal S-parameters of the circuit are simulated using ADS, and the input and output matching networks are designed, to match the base and collector impedances to the 50Ω source and load terminations. In the present case, single LC filter circuit on each side (input and output) are sufficient to match the circuit over the entire bandwidth. The circuit is re-optimized for better conversion gain, input and output return loss, and pass band flatness in the output

RF Synthesizer for Rb Atomic Clock   123

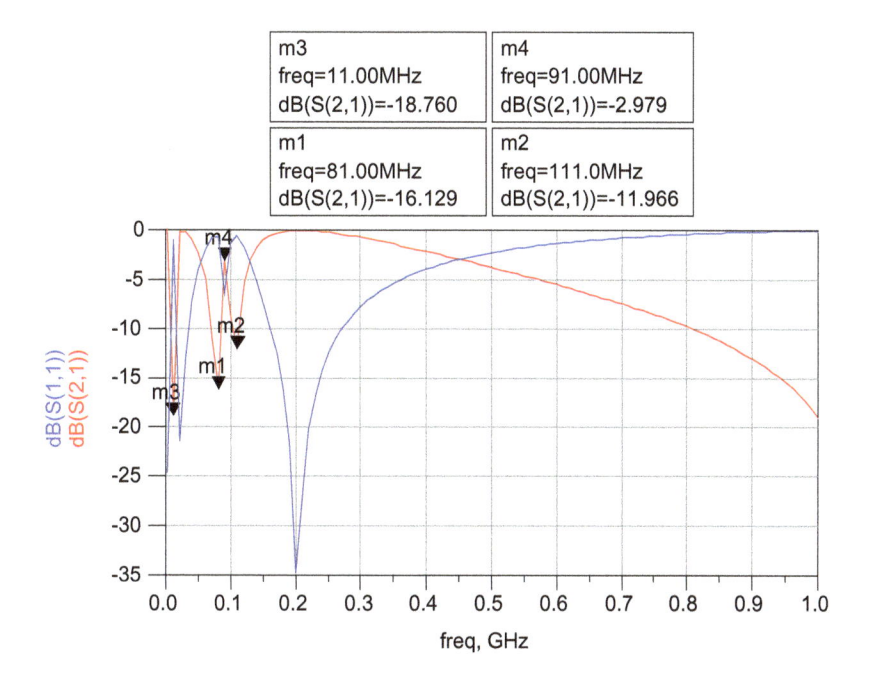

m3	m4
freq=11.00MHz	freq=91.00MHz
dB(S(2,1))=-18.760	dB(S(2,1))=-2.979

m1	m2
freq=81.00MHz	freq=111.0MHz
dB(S(2,1))=-16.129	dB(S(2,1))=-11.966

Fig. 5.6 Simulated idler network response.

after completing all the sub-networks design. The collector and base biases are fixed, based on the maximum output power at the required 9th harmonic, after simulating the transistor for an input power of 3 dBm, using the harmonic balance analysis of the Agilent ADS simulator. As shown in Fig. 5.7, the simulated results of multiplier for 3 dBm input, with idler network and optimized terminations, is giving a conversion loss of 3.2 dB.

5.4.4. *Experimental results*

The complete layout and the actual fabricated ×9 multiplier circuit is shown in Fig. 5.8. The circuit is designed and fabricated on a 0.8 mm FR4 substrate, with a dielectric constant 4.5. The proposed ×9 frequency multiplier circuit is fabricated on the 0.8 mm thick FR-4 laminate board using SMT devices. The circuit is biased at 3.0 V, 10 mA. The 10 MHz input is provided by Agilent E8257D PSG (250 KHz–40 GHz) analog signal generator. The output, fundamental

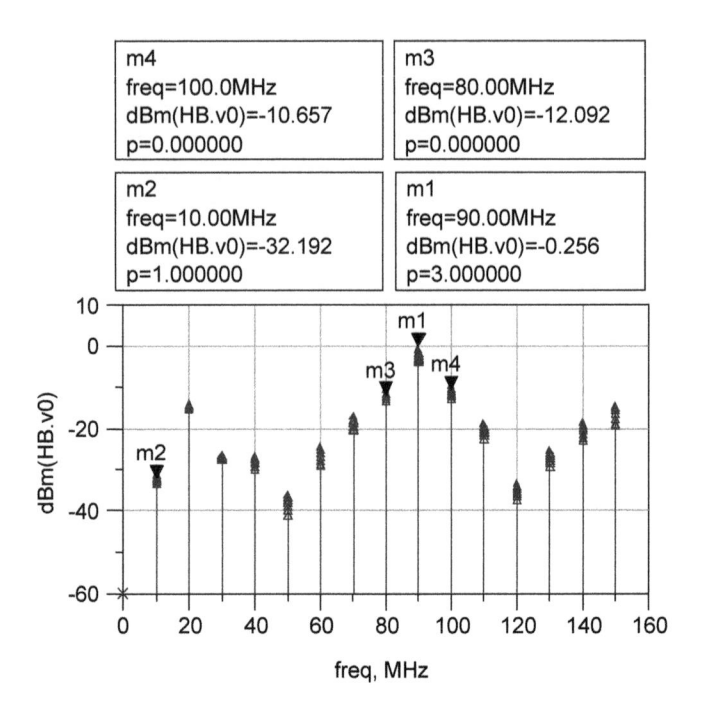

124 *Rubidium Atomic Clock: The Workhorse of Satellite Navigation*

```
m4                      m3
freq=100.0MHz           freq=80.00MHz
dBm(HB.v0)=-10.657      dBm(HB.v0)=-12.092
p=0.000000              p=0.000000

m2                      m1
freq=10.00MHz           freq=90.00MHz
dBm(HB.v0)=-32.192      dBm(HB.v0)=-0.256
p=1.000000              p=3.000000
```

Fig. 5.7 Simulated output spectrum of multiplier with idler network.

and harmonic frequencies are measured by Rohde & Schwarz FSU (20 Hz–43 GHz) spectrum analyser.

For a 3 dBm input power at 10 MHz, the output spectrum at desired 9th harmonic is measured. Figure 5.9 shows the wideband output spectrum of the ×9 frequency multiplier.

The rejection of 8th and 10th unwanted harmonic by more than 60 dB is achieved by putting a 90 MHz crystal filter at the output of multiplier. The measured output power and the conversion gain with respect to input power are shown in Fig. 5.10.

The measured conversion loss is 3.5 dB and it is in good match with the predicted simulated results. At 3 dBm input power, the output power at the 9th harmonic is fully saturated and the measured output power is −0.5 dBm with the 1 dB compression point. These measurements show good agreement with the simulations. Table 5.1 compares this work with other published work.

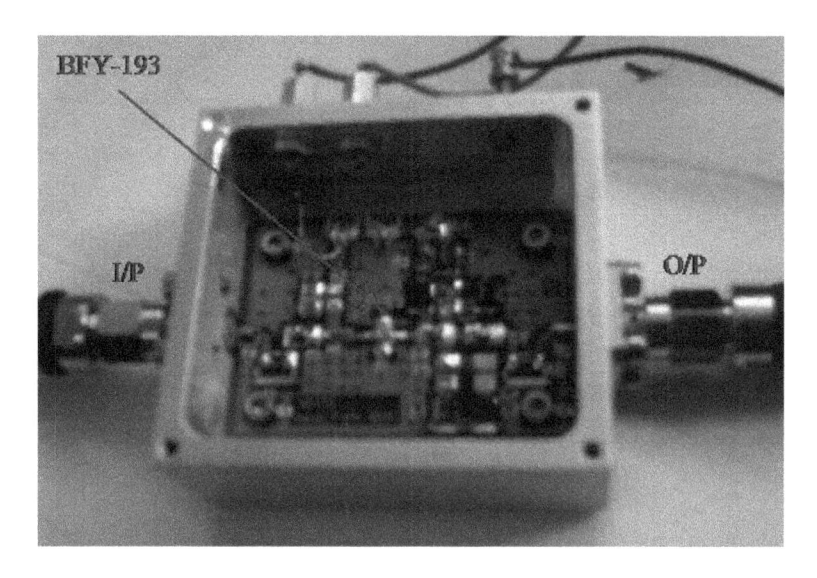

Fig. 5.8   Fabricated 10 MHz ×9 frequency multiplier.

The efficiency is calculated as the ratio of the output power at the 9th harmonic over the sum of the dissipated DC power and the input power at the fundamental frequency. The DC power supplied is 30 mW (3 V&10 mA) which gives dc-RF efficiency $\eta \approx 2.8\%$ at 90 MHz. The efficiency is written as

$$\eta = \frac{P_9}{P_{DC} + P_{In}}. \quad (5.6)$$

One likely explanation for the decrease in the efficiency is the resistor, connected in series with the base of the transistor to avoid stability related issue, which introduces loss into the circuit. Another cause of the loss is due to the idler lines, in the circuit and in the test fixture, used to test the packaged circuit. As the circuit is to be used in the space payload, certain de-rating guidelines are to be followed to improve the circuit reliability. Accordingly, the device cannot be operated at the maximum current and voltage ratings. Thus, by careful examination, the optimum operating point of the device is found to be 3 V, 10 mA. Figure 5.4 shows that the output waveform is in the pulse form and conduction is for less than half of

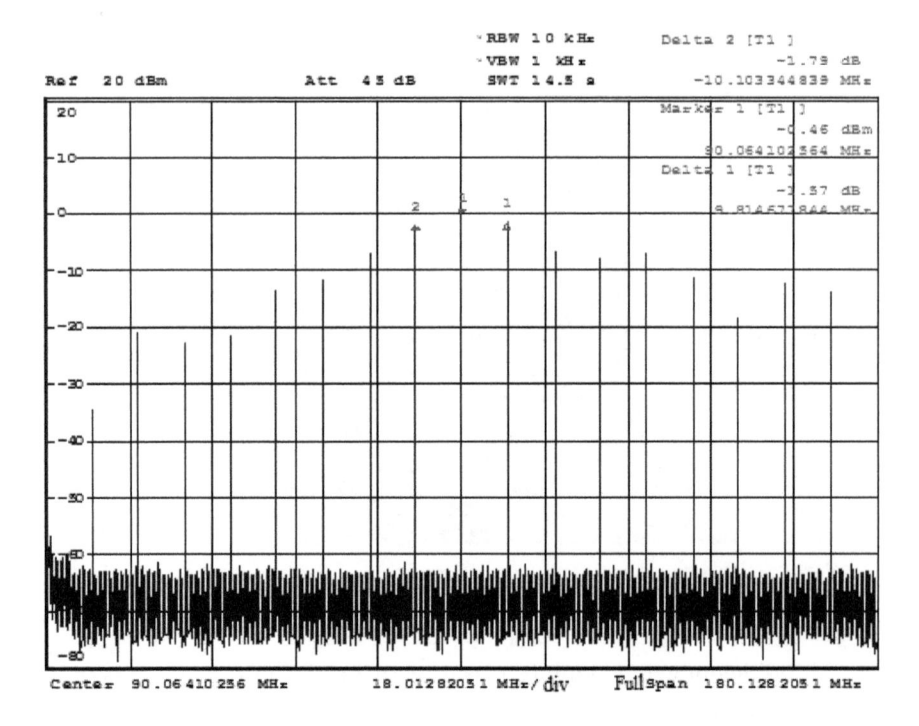

Fig. 5.9 Wideband output spectrum of the 10 MHz ×9 Multiplier.

the input duty cycle. It is also clear from Fig. 5.4 that the device is operated at a conduction angle of 170°, which is clearly not in match with the theoretical value of 28.8° as calculated from Fig. 5.3. The device is not operated in the extreme cut-off to keep the nth harmonic current levels within the de-rated values.

The phase noise is an important characteristic in frequency generation systems, and for that reason the evaluation of the output phase noise is required. A frequency multiplier is, in fact, a phase multiplier, which multiplies the phase deviation as well as the frequency of the input signal. This causes the phase noise degradation by 20log (N), where N is the multiplication order. In the case of a ×9 multiplier, it has the theoretical value 19.08 dB. The Rohde & Schwarz RF signal generator SMP-04 (10 MHz–40 GHz) is used for generating 10 MHz output. The output phase noise spectrum is measured by the PXA signal source analyser N9030A (3 Hz–26.5 GHz) from Agilent.

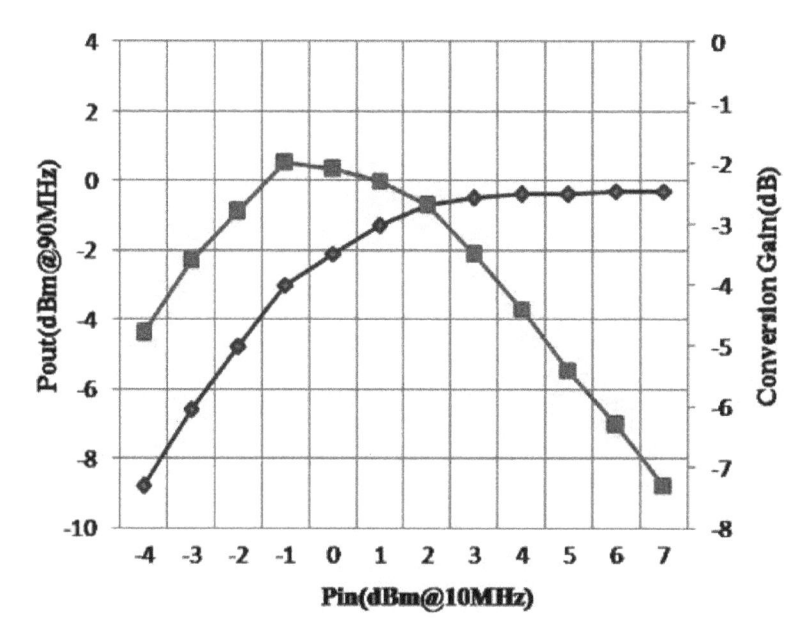

Fig. 5.10 Measured output power (Pout) and conversion gain with respect to input power (Pin).

To measure the output phase noise spectrum in presence of the other harmonic components, an external band pass crystal filter at the 9th harmonic of the input frequency is used between BJT frequency multiplier and the signal source analyser. For harmonic termination and better isolation, a 5 dB attenuator is used between the BJT frequency multiplier and the external band pass crystal filter. Figure 5.11 shows the measured output phase noise spectrum of the BJT, ×9 frequency multiplier for 0 dBm RF input power level.

The measured phase noise values at 10 MHz and 90 MHz carrier frequencies are shown in Table 5.2. From the measurement, it is observed that the phase noise degradation is in the range of 19 dB to 20 dB. The output phase noise is −101.646 dBc/Hz at 100 Hz offset, when the input power level is 0 dBm. The measured values at the output show that the degradation is 19.17 dB. The additive phase noise measurements at the fundamental and 9th harmonics

Table 5.1 Performance summary and comparisons of frequency multipliers.

Ref.	Technology	Multiplication Order	Output Frequency	Output Power	Input Power	Conversion Loss	Conversion Efficiency
8	SRD	10	12.5 GHz	−4 dbm	13 dbm	17 dB	≤2%
22	FET	3	56 GHz	−14 dbm	0 dbm	14 dB	<1%
24	GaAs FET	2	30 GHz	1 dBm	3 dBm	2 dB	<1%
26	BJT	2	4 GHz	−1 dBm	0 dBm	1 dBm	<1%
27	HEMT	2	6 GHz	−0.7 dBm	0 dBm	0.7 dB	2%
Our design	BJT	9	90 MHz	−0.5 dBm	3 dBm	3.5 dB	2.8%

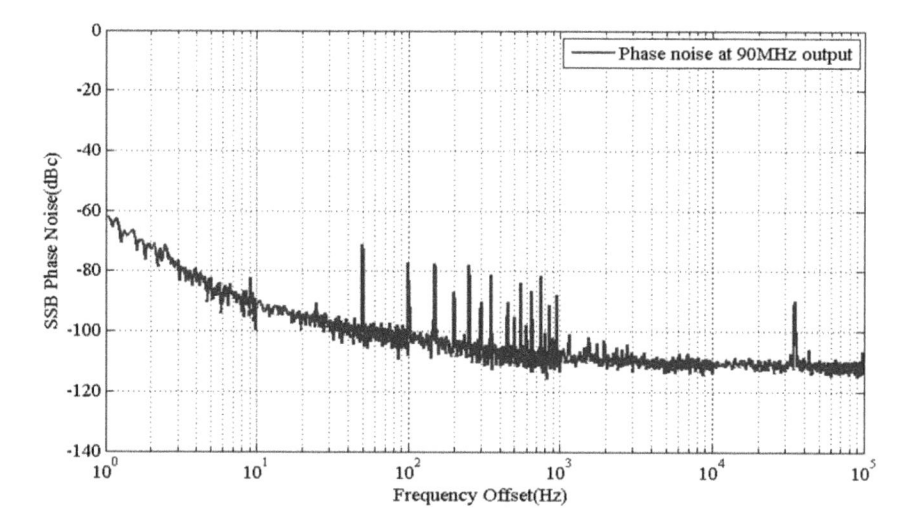

Fig. 5.11 Measured output phase noise spectrum of the BJT 10 × 9 frequency multiplier.

Table 5.2 Measured phase noise results at 10 MHz and 90 MHz carrier frequency.

Parameter	Phase Noise at 10 MHz Output	Phase Noise at 90 MHz Output
Offset frequency (Hz)	Measured Value (dBc/Hz)	Measured Value (dBc/Hz)
1	−81.175	−62.004
10	−110.126	−90.955
10^2	−120.817	−101.646
10^3	−130.152	−110.181
10^4	−132.198	−113.027
10^5	−132.215	−113.044

demonstrate true 20logN multiplication with an input phase noise of −120.817 dBc/Hz, at 100 Hz offset. The work demonstrates both theoretically and practically, that transistor frequency multiplier exhibits low phase noise for the high-order frequency multipliers. According to the requirement for a low phase noise of LO signal (less than -100 dBc/Hz at 100 Hz offset), the selection of a VCO with -120.817 dBc/Hz phase noise is a reasonable decision.

Summary

We conclude that the optimal output load, at the fundamental frequency is a short-circuited LC resonator or idler. The transistor model in Gummel–Poon study, used approximations. Therefore, the simulated data are only employed in the output termination effects. The developed experimental circuits based on our analysis, provide responses, which require tuning to the simulations. In pursuit of the optimal performance, the analysis contains both the simulated and the measured data. The measured results show the practicality of the designs and the accuracy of the models, which contain approximations. The accurate CAD techniques are utilized to develop bias, input power, input network and output network configurations for the optimum 9th harmonic response. As a result, the microwave $\times 9$ active frequency multiplier is designed and tested for an output frequency of 90 MHz. The designed multiplier gives a conversion loss of 3 dB for input power level of 3 dBm with 2.8% efficiency. The 8th and 10th harmonic rejection is better than 60 dBc. The results are quite excellent for an $\times 9$ order multiplier. The performance of this active multiplier circuit shows its feasibility for the replacement of SRD multipliers in LO chains, and thus eliminating the need of pre-amplification and post-amplification stages. The measured phase noise degradation in the range of 19 dB to 20 dB, indicates that this circuit is a good alternative in reaching the low phase noise requirements for a LO chain.

5.5. Design and simulation of 360 MHz × 19 SRD multiplier

5.5.1. *Introduction*

Step recovery diodes (SRDs), strongly non-linear two terminal devices, are used as comb generator, wave formers [136] and in hybrid local oscillators, especially where low phase noise is required, (i.e., in terrestrial communications, satellite communications, TVRO, low cost UWB transmitters [137–139], mobile communications) As a frequency multiplier, the SRD is used for example, in millimetre wave link radios. Another typical application of the SRD is as a comb generator in microwave and millimetre wave samplers. It is used in frequency counters, sampling scopes, phase locked synthesizers, and network analyzers [140–142]. One of the most outstanding characteristics of the step recovery diode (SRD) is the high conversion efficiency with high frequency multiplication order. It provides a method for generating power at high frequencies by using a low cost oscillator. Input frequencies of SRD may be down to 10 MHz and output frequencies up to 94 GHz [136, 143].

There are several variety of the frequency multipliers, designed by means of comb generator [144–145], which produce a set (or "comb") of discrete harmonically related tones in the frequency domain corresponding to a periodic waveform in the time domain [146]. Several analyses of harmonic generation circuits, using the SRD, are restricted in applications by their dependence on the knowledge of the waveforms of currents, flowing through the diode [147–150]. In the past, designs of SRD frequency multipliers were mainly based on the method of Hamilton and Hall [151]. The circuit solving algorithm, used by simulators, is based on the Newton–Rapson algorithm, so cannot be used directly in commercial circuit simulators. Only one conduction angle per cycle has often been analysed in the past, under the assumption that the amplitude of the nth harmonic was small. Recently, more than one conduction angle per cycle has been observed in an analog simulation study [152–153]. A new SRD model, more accurate because of considering the voltage ramp during the transition process, has been realized using the CAD. It can be used directly in commercial circuit simulators [154]. A model of the SRD is created based on the extracted diode parameters, and the delay-line SRD impulse generator, described by G. D. Cormack [155]. The circuit is very typical and helpful in simplifying the analysis. In 1983, a computer simulation model of a conventional p-n junction diode was proposed by Goldman [156], and characterized by experiments. However, as we know, a SRD is essentially different from a conventional p-n junction diode and due to its very strong non-linearity, its characteristics cannot be modelled very well for all the design conditions. With the ever rising demand of integrated circuits, more accurate designs are required for developing the circuits for the conditions where tuning is not possible. Of course, there are other reasons which influence the design of the circuit. The microstrip circuits have small size, low weight, high reliability. In parallel, the development of computer aided design tools provide us with the possibility of simulation and optimization of the circuits, to achieve an accurate design and optimum performance.

The purpose of this section is to present some of the critical performance characteristics of SRD diode, as these represent the features of multiplier design for the stable operation and the maximum efficiency. In this section, a model of the SRD circuit is developed on CAD. It can be used in any circuit simulators. A modified basic model of the SRD is established, followed by the applications of this model to analyse the comb generators and frequency multipliers. In order to improve the efficiency of the CAD of SRD frequency multipliers, we investigate the modelling of the diode and features of SRD frequency multipliers. The step-recovery-diode multipliers are analysed and optimized using the harmonic balance methods [157-158]. The results show that abating the non-linearity of the diode model to some extent for optimizing the circuit, does not radically change the characteristics of the circuit. Due to the strong non-linearity of the diode and the generation of a large number of harmonics, harmonic balance analysis of the SRD frequency multiplier can be difficult and the possible instability make convergence precarious. In addition, to choose an advance simulator for good convergence, more efforts should be made for the correct solution for this type of circuit. However, the designs based on this model have very limited accuracy, and high performances of actual circuits may usually be achieved by experimental adjustments.

A systematic design of a 360 MHz ×19 comb generator, is given as an example. The detailed analysis is done for the input-output matching circuits, impulse generator circuit and achieving the minimum power of −26 dBm at 6.840 GHz and the pulse width less than 1 ns. The comb generator is fully characterized by varying the input power over the range of 20 dBm ± 5 dB and the corresponding output power variation at 19th harmonics is also recorded. The variation of noise floor close to the output frequency (carrier ± 1 MHz) is characterized by changing the input power. Then, the effect of inserting transmission line between SRD and the filter circuit is studied. The repetitive measurements show that this comb generator has magnitude uncertainty of ± 0.3 dB at (25±1)°C room temperature, at rated 21 dBm input excitation, for which the output power fluctuation is less than 0.8 dBm.

This description of the multipliers is organized as follows. Section 5.5.2 discusses a SRD model and verified through simulation and experiments and Sec. 5.5.3 analyses the design detail in ADS software and harmonic balance simulation. The experimental results are discussed in Sec. 5.5.4 and the summarized in Sec. 5.5.5.

5.5.2. Functional model of SRD multiplier

The step recovery diode (SRD), which is also called the snap-off diode or charge storage diode, was first discussed in the early 1950s. The step recovery diode (SRD), or "snap diode", has an unusual operating characteristic to store a large amount of charge in forward biased condition, and conduction in the reverse biased state, until all the stored charge is recovered by a negative current. Then a very abrupt turn-off happens that ends in sub-nanosecond transitions (<1 ns) [159]. So an ideal step recovery diode functions as a switch from a high impedance state to a low impedance state. Which corresponds to a small reverse bias capacitor (depletion C_r) and a large forward bias capacitor (diffusion C_f), with zero switching time between states. The limiting factor in switching from C_f to C_r is the rapidity with which charge can be extracted from the i layer. The voltage ramp begins when the forward bias is reduced, after that there is a fast transition and finally a rounding off, until all the stored charge is swept out from the i layer. Then the non-conducting state continues as the voltage becomes more negative. Thus, the conducting state is transformed to the non-conducting state through a process of turn-off.

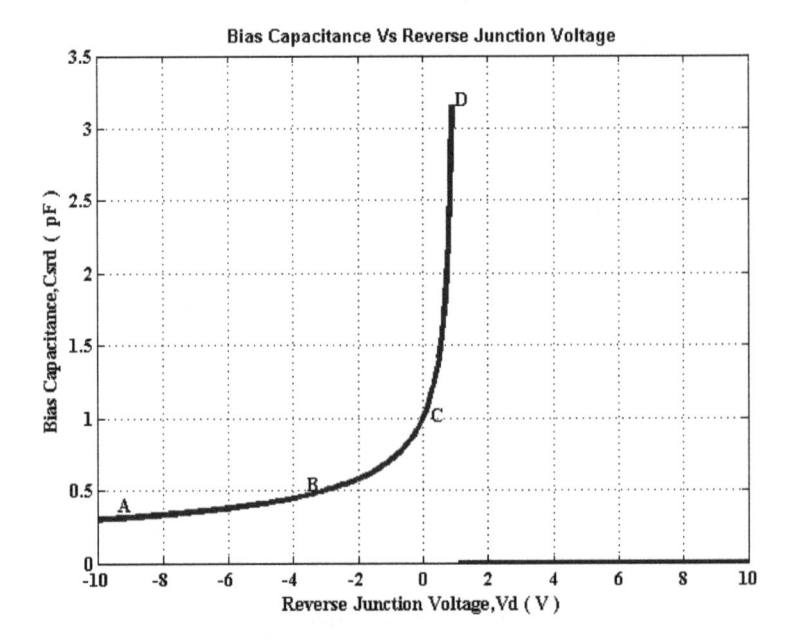

134   Rubidium Atomic Clock: The Workhorse of Satellite Navigation

Fig. 5.12   The C-V curve of SRD.

Accordingly, lines CD and AB represent the conducting and nonconducting bias capacitances, respectively in Fig. 5.12. The BC section is a parabolic curve representing the turn-off process. When SRD is reverse biased, capacitance is nearly constant and also is very small. However, when SRD turns into forward biased, capacitance becomes quite large rapidly, an abrupt change, which is just an intrinsic factor of attaining high harmonics together with high efficiency. The transition process can be described by a parabolic function, which is determined by the two state capacitance [154]. The characteristic of the diode shows that the smaller is the forward capacitance, the weaker is the non-linearity of the model. This highly non-linear capacitance characteristic is also accompanied by a highly non-linear shunt resistance, which depends on the forward bias voltage and RF input power. Therefore, the degree of abatement depends on the changes made in C_f and the operation condition of the diode, Fig. 5.13.

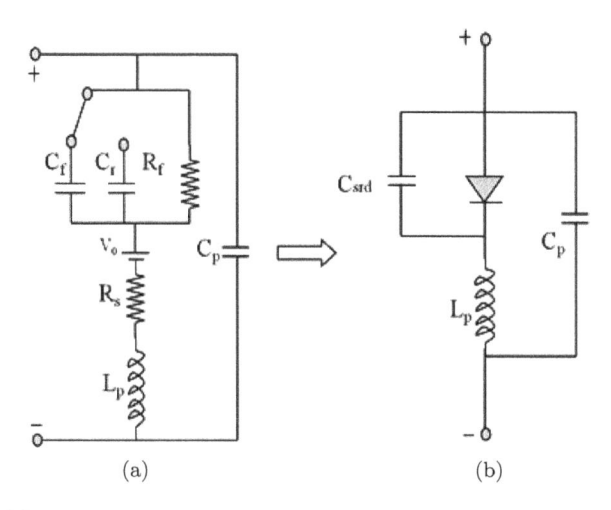

Fig. 5.13 (a) Conventional model of SRD and (b) Fully functional SRD model for CAD simulation.

The functional aspects of the SRD are based on the model proposed by J. Zhang and A. V. Raisanen [160–161]. In this model, the diode series resistance R_S is internal to the PN diode circuit. While in the referenced publications, it is external to the diode and placed in series with the package inductance L_P. The SRD capacitance is the centrepiece of the model. It is a simple piece-wise linear model consisting of three line segments: a small capacitance when the diode is reverse biased ($V_d < 0$), a large capacitance when the diode is forward biased ($V_d > F_C.V_j$) and a linear ramp ($0 < V_d < F_C.V_j$). Specifically,

$$C_{srd} = C_r \quad \text{for } V_d < 0$$
$$= C_r + \frac{\tau/R_f - C_r}{F_C.V_j} V_d \quad \text{for } 0 < V_d < F_C.V_j$$
$$= \frac{\tau}{R_f} \quad \text{for } V_d > F_C.V_j,$$

where V_d is the voltage across the capacitance and other model parameters C_r, τ, R_f, F_C and V_j are the reverse bias capacitance, minority carrier lifetime, forward bias resistance, forward bias depletion capacitance coefficient, junction potential respectively. When the SRD is forward biased, the p-n Junction acts like a dynamic or non-linear resistor. The relationship of C_f, RF and the minority carrier lifetime can be related, as given by Kotzebue [162],

$$\tau = R_F C f,$$

where 'τ' is the minority carrier lifetime of the SRD provided by the manufacturer. The assumption is that the value of the forward capacitance $C_f = \tau/RF$ is larger than the reverse capacitance C_r. If it is not, it is snapped at C_r, thus making the overall capacitance linear. If both parameters C_r and τ are set to zero, the SRD capacitance C_{srd} is effectively, removed from the model. The standard p-n diode junction and diffusion capacitances are available and can be included if desired, by setting the corresponding parameters C_{io} (zero bias junction capacitance) given by the manufacturer and T_t (transit time) to non-zero values. For a realistic model, C_f is much larger than C_r, thus making the overall model highly non-linear. In addition to the two constant capacitors, the series parasitic resistance, the package inductance and capacitance should be taken into consideration. The equivalent circuit of an SRD including all these factors is shown in Fig. 5.13(a) and a fully functional SRD diode model for CAD simulation is given in Fig. 5.13(b).

5.5.3. Simulation and design methodology

The model by Moll and Hamilton [163] is a base for the design of impulse circuit, established by Zhang and Raisanen [154]. The series parasitic resistance, the package inductance and capacitance have some influence on the model, in addition to the two constant capacitors. On the other hand, we also need to select proper SRD for the simulation. In the simulation of circuits, good initial conditions are sometimes critical for convergence. Besides that, they often help the simulator use less computational resources (time and memory) to find the correct solution. Normally, there is a way to set initial conditions in the simulator. Using a simplified circuit is one way to find good initial conditions. Another process, for obtaining convergence at the time of simulation, is to start with values of certain components, that work and then to move toward the desired values. Several important parameters of SRD are: the step recovery time T_t, the minority carrier lifetime τ and reverse junction capacitor C_r. The values of the parameters for setting the initial condition in the simulation are as follows:

(a) step recovery time $T_t \leq 1/f_0$, ($1/f_0$ is output signal period);
(b) the minority carrier lifetime $\tau \rangle \rangle 1/2\pi f_{in} (\omega_{in} \tau > 10$ is adequate for most purposes);
(c) reverse junction capacitor $C_r <= 1/2f_o X$, where $10 < X < 20$; where X is approximate impedance level assumes 50 Ω system;
(d) reverse voltage is higher than impulse amplitude.

Based on those above, we choose DH543-62A produced by TEMEX, whose $T_t = 90 - 140ps$, $\tau = 20ns$, $C_r = 1pF$ and the minimum of reverse voltage is 30 V.

From the characteristic of SRD, we know that the smaller the forward bias capacitance, the less non-linear is the diode model. Thus, first we can reduce the non-linearity of the diode model by changing the forward bias capacitance to a smaller value. The reduction of the non-linearity of the model may possibly, help the simulator to find a solution, which can be used to obtain the desired final solution efficiently. It can be seen that the forward biased resistance turns into a very small and nearly constant value right after the diode is forward biased. In the case of the diode under test, the value of the forward biased resistance is of the order of 1 ohm, as given in the data sheet, which corresponds to a forward biased capacitance of 35 nF. However, the depletion capacitance of this diode, which is used as the reversed bias capacitance C_r in our model, is about 1 pF and comparing with the calculated C_r, shows the very strong non-linearity of the diode model. This model in circuit simulators definitely costs a lot computational resources and most probably, causes convergence problems. By abating the non-linearity of the model of the diode to an appropriate extent, the simulation and optimization of SRD frequency multipliers can be carried out faster.

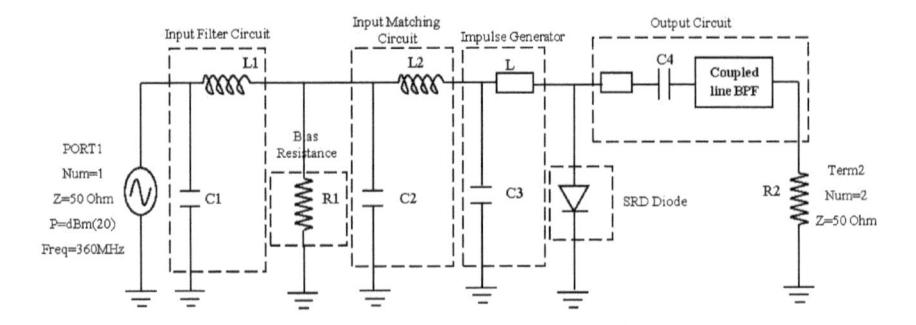

Fig. 5.14   The schematic circuit of an SRD.

Based on the modified SRD model, the simulation is directly done with the Advance Design System (ADS). A systematic schematic diagram of the SRD frequency multiplier circuit and ADS-created simulated circuit model are shown in Fig. 5.14.

The SRD frequency multiplier can be divided into three parts: the impulse generator, input matching circuit and output circuit of comb generator. The impulse generator converts the energy in each input cycle into narrow large amplitude voltage pulse, occurring one per input cycle. The input matching circuit provides a match to the 50 Ω source impedance at the 360 MHz, the generator driven frequency. The output resonant circuit converts the impulse into a damped ringing waveform at the desired output frequency. The output circuit of the comb generator, which consists of transmission line, coupling capacitor and the output band pass coupled line filter, are used to filter out the desired harmonic from the output circuit of the frequency multiplier. On account of the wide-band output spectrum for 1 GHz ∼ 10 GHz of the output port, we cannot go along to match, for any single frequency from the point of view of the energy. If energy concentrates at a certain frequency point, then the energy of other frequencies becomes very small. The micro-strip line can be regarded as invariable over very wide frequency band from low frequency to high frequency. Therefore, we employ micro-strip line, connecting to 50 Ω interface system at the output port. However, when the frequency is higher than 5 GHz, TEM wave produces dispersion, so we need to find a substrate, which satisfies the above conditions. We find that a thin substrate with low relative dielectric constant is able to meet the requirement of the ADS simulation. Consequently, we utilize alumina substrate for the output port, whose relative dielectric constant εr and substrate thickness are is 9.9 and 25 mil respectively.

We designed a 19 × 360 MHz SRD micro-strip frequency multiplier. The harmonic-balance simulation is completed with ADS using 35 harmonics at an input frequency of 360 MHz and power +21 dBm. At first, the circuit is optimized for the high conversion efficiency with the driving inductor, realized by a micro-strip line L and RF capacitor C_3. The detailed calculation [154, 162, 164] of a straight micro-strip line inductor is simple and can be obtained by using Hamilton Hall's method [151, 162] to tune it at input frequency, shown in Fig. 5.14. The harmonic-balance analysis with swept RF power, are carried out for the two values of C3 (i.e., 3.6 pF and 3.9 pF) and compared with the experiment. Then the circuit is optimized and refined experimentally, for high conversion efficiency with C3, equal to 3.6 pF. The optimum source impedance can be defined as the impedance that gives the maximum conversion efficiency at the desired harmonic. Once this optimum source impedance is known, an impedance matching circuit can be designed by using ADS simulation software, to match the diode to a 50 Ω source. For the best efficiency, it is necessary to match the relatively low impedance of the SRD circuit to 50 Ω at the input frequency. The additional lumped components, L2 and C2 provide a match to the 50 Ω source impedance at the 360 MHz generator driven frequency, and the external bias is replaced by the self-bias resistor R1. The quarter wave resonant transmission line at 6.8 GHz is one way to boost the level of the desired harmonics, above the other harmonics, before the additional filtering. The micro-strip line length is calculated according to the highest frequency, namely 6.8 GHz, and its length is less than a quarter of the wave length. The line is terminated with a resistance, chosen so that the loaded Q of the resonant line is approximately $(\pi/4)$ N, where N is the ratio of desired output frequency to input frequency (N = 19 for this design). The output of the resonant line

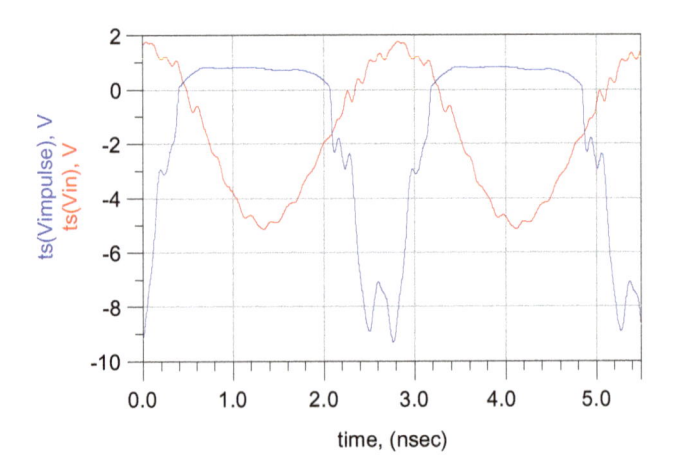

Fig. 5.15 The input waveform and that of the impulse at the diode/transmission line node.

for this value of Q is a damped sine wave, that spans one input cycle. However, a piece of 50 Ω transmission line for the output port is still not the best, so Q is adjusted by varying the degree of coupling to the 50 Ω loads, with a small series capacitor, which can block the DC and partially reflect the RF output. The length of line is adjusted so that the sinusoid rings with a period of 120 ps. It also provides us with more design freedom for the optimization of the circuit. Figure 5.15 (Red curve) notes the voltage waveforms at the input of the comb generator and at the junction of L1, L2, and C1. Note that the impulse has not been totally filtered at this point. Blue curve shows the impulse generated into 50 ohm load without the output filter.

By varying parameters such as junction capacitance, reverse recovery time, and parasitic inductance; different pulse shapes can be optimized. Note that the circuit layout in Fig. 5.14 is basically a comb generator and it is going to be used as a part of the frequency multiplier. This part of the circuit can be optimized either at this stage or together with the output filter circuit of the frequency multiplier at a later stage. Thus, the transmission line and the capacitor can be optimized to give the maximum conversion efficiency, nevertheless, the output signal should be filtered further.

5.5.4. Design of 6.8 GHz coupled line filter

The design of output circuit of the frequency multiplier consists mainly, of the design of a band-pass filter. The bandwidth of the filter should be determined according to the specific application. Normally, narrow band coupled line band-pass filter is used. Whereas wide band-pass filter could also be employed for a particular application [165]. The design method can be found in other literature [157, 166]. Here a micro-strip coupled line band-pass filter is designed and simulated using ADS. Once the designs of the comb generator and the band pass filter are accomplished, they can be put together to form the frequency multiplier. However, since the output of the comb generator is very rich in harmonics and the desired harmonics could be a very high order harmonics, the effects of all those unwanted harmonics can not readily be neglected. Apparently, the filter can extract the desired harmonics from the comb generator, and drop the unwanted harmonics. The comparison of simulation results with and experiment is shown in Fig. 5.16.

The band-pass filter are designed, fabricated and tested on alumina substrate, using MIC with a relative dielectric constant of 9.9. The simulated and measured results are shown in Fig. 5.17. The measured centre frequency is 6.834 GHz and the measured bandwidth is 600 MHz. The insertion loss is less than 1 dB and return loss is 23 dB. The circuit-simulation shows good agreement with experimental results. The electrical length, line impedance, and coupling capacitor all affect the shape and frequency response of the final signal. The goal of varying these component values is to maximize the power available in the nth comb. Finally, the harmonic balance simulation results at the SRD output and cascading SRD with filter output are shown Fig. 5.17, and they provide a look at the output spectrum centred around 6.834 GHz ($\times$19 of input frequency), when the input power is 21 dBm. It can be seen that the simulated results are relatively satisfying, the minimum output power reach -25 dBm. Nevertheless, it should be pointed out that the simulation results are very ideal. Therefore, the input matching circuit and the output

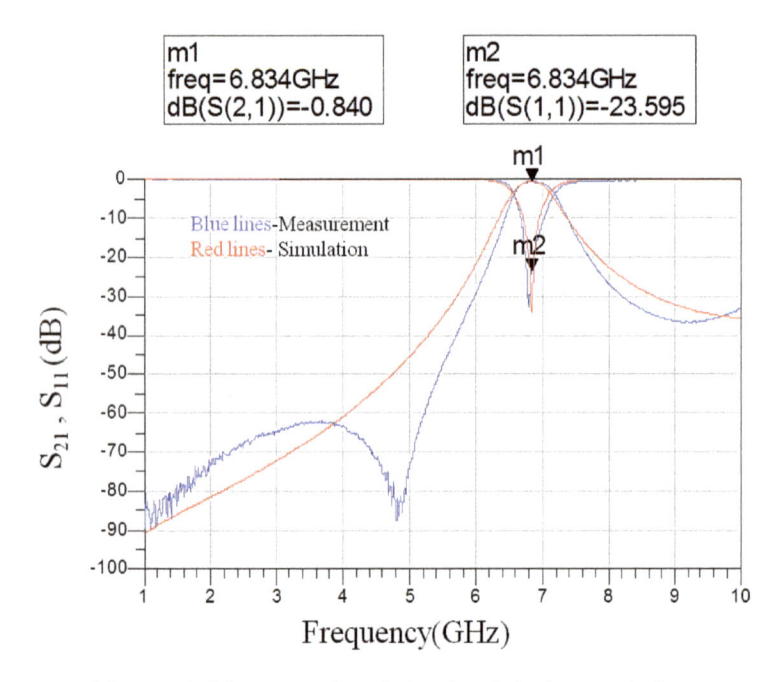

Fig. 5.16 Measured (blue curves) and simulated (red curves) S-parameters of coupled line BPF.

circuit consisting of a transmission line, a capacitor and a band pass filter should be re-optimized to get the optimum performance.

5.5.5. *Experimental results*

A 19 × 360 MHz SRD frequency multiplier is realized with the microstrip circuits and TEMEX made DH543-62 ceramic packaged SRD diode. The input circuit is used for impedance matching between the 50 Ω source and the low impedance of the diode. The output circuit consists of a section of transmission line, a capacitor, and a separate micro-strip band pass filter. At last, an impulse circuit is fabricated and the experiment is carried out with a 360 MHz input sinusoidal signal at the power level of 21 dBm. The whole pulse generator circuit fabricated on 25 mil alumina substrate with its dielectric constant of 9.9, the overall dimension of the assembly is 19 mm × 10 mm × 0.6 mm.

RF Synthesizer for Rb Atomic Clock 143

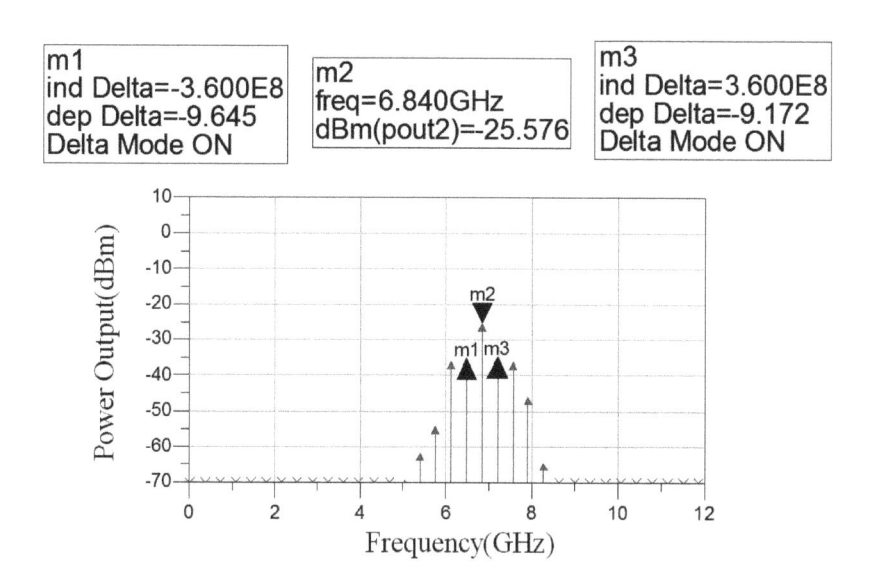

```
m1
ind Delta=-3.600E8
dep Delta=-9.645
Delta Mode ON
```

```
m2
freq=6.840GHz
dBm(pout2)=-25.576
```

```
m3
ind Delta=3.600E8
dep Delta=-9.172
Delta Mode ON
```

Fig. 5.17 Simulated power spectrum at filter output.

The substrate thickness of the comb generator is chosen to match the ceramic package. The band pass filter could also be integrated on the same substrate as the comb generator. However, a separate band pass filter, which has more freedom in choosing substrates, and which provides the flexibility of experimentation, is used. The high performances of actual circuits are usually achieved by experimental adjustments. In the process of adjustments, we find that input power and some components of matching circuits have influence on the output power to a certain extent. Using Agilent E8257D PSG analog signal generator (250 KHz–40 GHz) as the Source, and Rohde & Schwarz FSU spectrum analyser (20 Hz–43 GHz), the power at SRD output is measured and the experimental results are shown in Fig. 5.18

The experiment is carried out by changing the input power (20 dBm ± 5 dB) and recording the corresponding output power at the 19th harmonic and characterizing the noise floor close to the output frequency (6.84 GHz ± 1 MHz). The measurement results show that above 21 dBm of input power level, the output power at 19th harmonic is almost saturated and signal-to-noise ratio is also improved (>10 dB) at higher input power (>21 dBm), in Fig. 5.19.

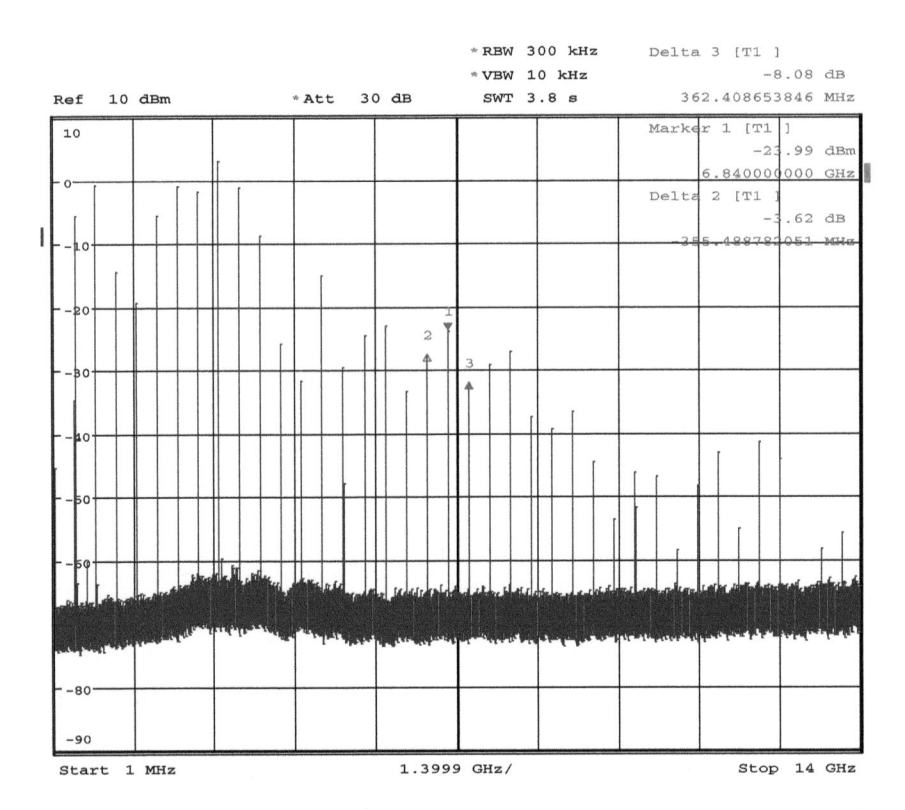

144 Rubidium Atomic Clock: The Workhorse of Satellite Navigation

Fig. 5.18 Power spectrum at SRD output.

Both the simulation and the experiment show that the circuit is quite sensitive to the length of the transmission line, between the diode circuit and output filter circuit, which is closely related to the mounting structure of the diode. For better modelling of the hybrid SRD frequency multiplier, we use the numerical electromagnetic simulation technique, which analyses the effects of the diode mounting structure. The 18th, 19th and 20th harmonics are characterized by inserting various transmission line lengths between SRD multiplier and the filter cascade combination. The measurement results are shown in Fig. 5.20.

The comparison between simulated and measured performances is given in Table 5.3. The lower values in the experimental results vis-á-vis the calculated values are due to the loss in the circuit.

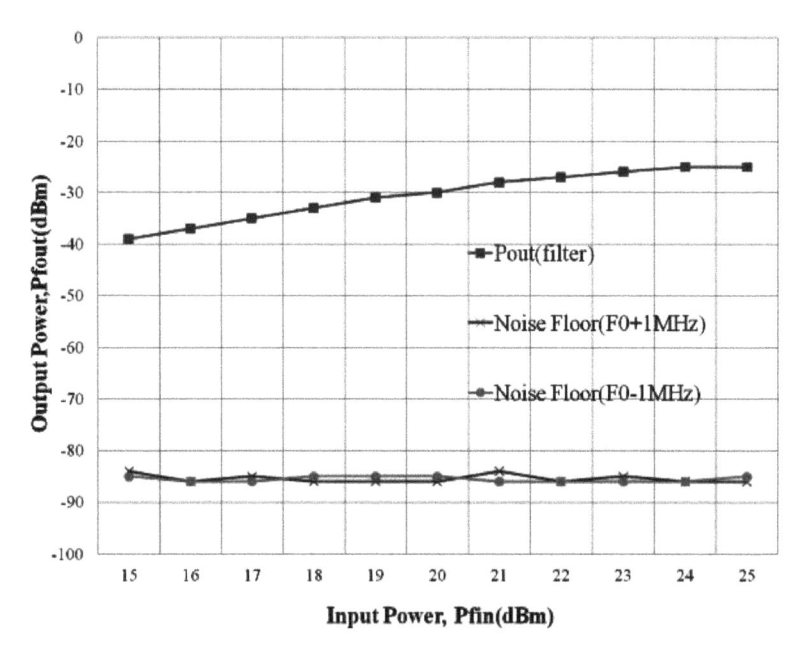

Fig. 5.19 Experimental results (a) output power vs input power (b) noise floor vs input power.

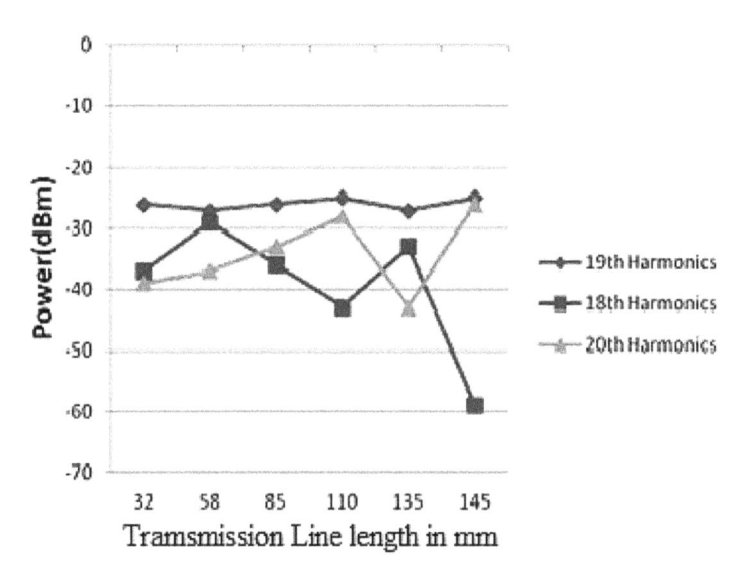

Fig. 5.20 Output power vs transmission line length at 18th, 19th and 20th harmonics.

Table 5.3 Simulation vs measured results of SRD multiplier circuit.

Results	Input Power	Harmonics		
SRD only	P_{In}	P_{out}@6.48 GHz	P_{out}@6.84 GHz	P_{out}@7.20 GHz
Simulation	21 dBm	−25.21 dBm	−22.67 dBm	−27.35 dBm
Measured	21 dBm	−27.61 dBm	−23.99 dBm	−32.07 dBm
SRD ÷ Filter				
Simulation	21 dBm	−35.21	−25.57	−34.74
Measured	21 dBm	−35.25	−26.28	−38.88

Fig. 5.21 The actual circuit board of SRD and BPF.

The photograph of impulse generator circuit with the filter circuit is shown in Fig. 5.21.

5.5.6. Results

The CAD of SRD frequency multipliers can be made more efficient and easier to accomplish, by reducing the simulated non-linearity of the SRD model. The simple generic ADS model for SRD and all passive components of a 360 MHz to 6.84 GHz multiplier predict actual circuit operation accurately, for an output frequency in the desired microwave range. A parameter-extracted model of SRD is firstly, implemented for DH543-62A, and its feasibility is verified by a test circuit. The variations in the transmission line length between SRD and the output filter circuit, that influence the 18th, 19th and 20th harmonic contents, are studied with the simulation and the experimental method. The experiment shows that the simulation results can be used to guide the SRD frequency multiplier design, achieving a good signal to noise ratio. In general, this design process may also be applied to CAD of circuits, using other highly non-linear devices. By using these circuits, we may also easily fabricate frequency multipliers. We match the output with the desired single frequency, and then introduce a band-pass filter. The advantages of this approach may include more flexible resolution setup, little hardware modification requirement and better spectrum energy utilization.

5.6. Design of two stage 6.8 GHz amplifier

5.6.1. *Introduction*

In the work reported here, a 6.8 GHz two-stage MIC amplifier is developed with a small signal gain of 24 dB. The amplifier is based on GaAs HEMT technology, with $f_T = 12$ GHz and $f_{max} = 20$ GHz. In this design a cascade configuration in combination with MIC technology is adopted.

5.6.2. *Design concept of two-stage amplifier*

As shown in Fig. 5.22, the power amplifier topology is a 2-stage cascaded design with an interstate complex conjugate matching techniques between the two amplifiers, both using a pHEMT. The 2-stage design is used, since a single stage is unable to meet the gain and power specification at the design frequency. Finally, we achieve performance for the two stage power amplifier with good linearity, 20% drain efficiency and an output power up to 10 dBm, with about 24 dB power gain over the bandwidth for -5 dBm input power.

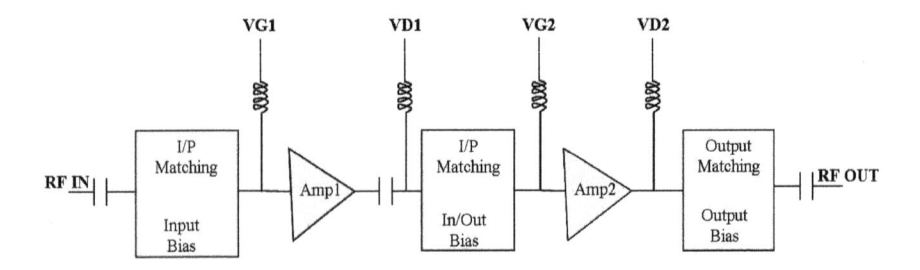

148   Rubidium Atomic Clock: The Workhorse of Satellite Navigation

Fig. 5.22   Schematic of two-stage amplifier.

5.6.2.1. Bias network

The RF/Microwave transistors/FET require the circuit to set the correct bias conditions for the desired RF performance at the operating frequenct 6.834 GHz. The role of the bias network is to block unwanted DC and RF from either signal and DC sources, as both DC and RF are present at the device. What is required is a low DC resistance but a high RF resistance to ensure that the RF circuit is not loaded, and RF signals do not enter the supply lines. In order to block DC, a low value capacitor (∼1 pF capacitor) or coupled line 6.8 GHz filter is used at the input and output. Improper design of a DC bias network may lead to undesired oscillations, and RF signal leaking to the DC supply. Hence, it is worthwhile to evaluate the bias network before designing the two stage amplifier. The objective of the bias circuit design is to minimize the insertion loss, while obtaining the high RF isolation to the DC source. The design of bias network uses a high impedance quarter wavelength transformer, with either a radial stub [167] or chip capacitors. In order to ensure open circuit condition at the operating frequency, one probable and commonly used method is to place an open radial stub immediately after $\lambda/4$ high impedance bias line [169]. The bias lines are designed with quarter-wave short stubs at a centre frequency of 6.8 GHz. The bypass open radial stub short circuit these quarter-wave lines in the in-band frequencies, and isolate the rest of the amplifier from the power supply. In addition, TFRs (Thin film resistor) and capacitors are included in

RF Synthesizer for Rb Atomic Clock 149

the bias networks for improving the stability. This helps to achieve proper isolation at desired RF frequency, no matter what component is added after $\lambda/4$ long bias line. The principal behind this design is that the open radial stub is transformed into short circuit at the tee. The 90° transmission line then transforms from short circuit to open circuit, at the gate of the device. This ensures that at the frequency of operation, the bias network is seen as an open circuit. The bias network consists of a quarter-wavelength transformer at 6.8 GHz and a radial stub. The length of the stub is slightly shorter than quarter-wavelength and the angle subtended by the stub is 45°. The circuit is simulated in the ADS 2009. Figure 5.23 shows, that the insertion loss of the bias circuit is less than 0.1 dB in the RF frequency range 5.7 to 7.7 GHz, with a return loss of -20 dB. A coupled line filter is designed in ADS software.

5.6.2.2. *Matching network*

The circuit design task is initiated using a cascaded part, with the objective of obtaining superior gain and good stability of the

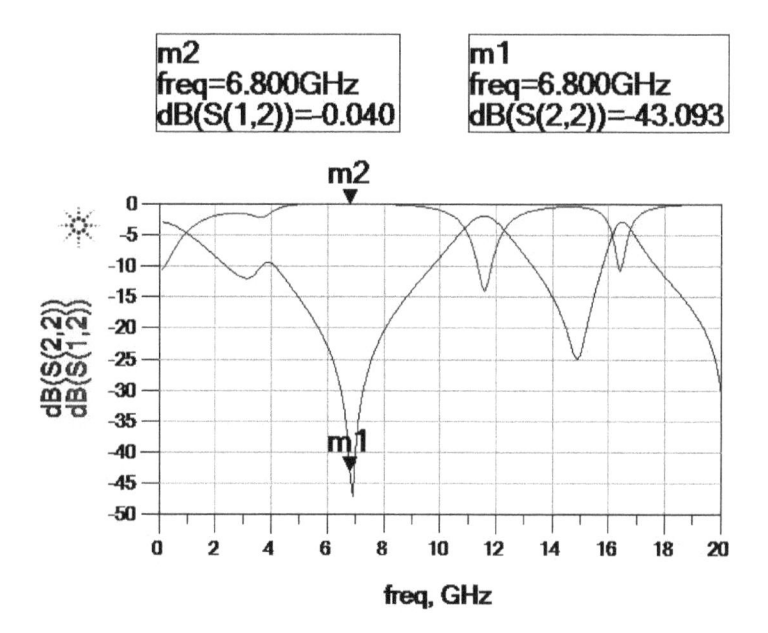

Fig. 5.23 ADS simulation insertion loss, return loss of bias tee circuit.

operating frequencies. The input and output are separately matched for 50Ω impedance, using open stub networks. The inter-stage matching is optimized for the maximum gain with a capacitor, blocking the low frequency signal, and thus enhancing the stability of the amplifier. After the design for single stage amplifier is completed, the next step is to combine the two single stages together with the inter-stage connection. The output impedance of first stage and the input impedance of second stage are analysed in ADS simulation software. The 1st stage has no matching on the output, because S_{22} of 1st stage is now modified by adding the input matching circuit of the second stage. We require a reasonable output return loss, for that matching to the conjugate of modified S_{22} of the 1st stage is done. This is because S_{22} is looking into the device and the conjugate is looking towards the matching circuit. The conjugate matching techniques are used to design the inter-stage matching network.

The intermediate matching section should transfer the impedance from the modified S_{22} to the S_{11} of the 2nd stage HEMT (CFY-67). The inter-stage matching network consisting of a capacitor and two open radial stub, each from bias network in pre- and post-stages. It really helps us to save the lumped components and minimize the circuit size. The capacitor in inter-stage connection also provide DC block for both pre- and post-stages. The 2nd inter-stage matching network is optimized to obtain the best conjugate matching between the optimum source impedance and optimum load impedance, as well as port to port isolation.

The input and 1st inter-stage networks are optimized to obtain small-signal gain flatness for the complete amplifier gain performance and decent input port return loss. After the optimization, the complete circuit is simulated in the ADS.

5.6.2.3. Optimization of multiplier design

Figure 5.24 shows the layout of the two stage amplifier module designed in the ADS. With operating frequency in the microwave band, the cavity resonance effects become a common problem, when the standing wave ratio (SWR) peaks at certain points. In the design, more isolation walls and blocks are added.

RF Synthesizer for Rb Atomic Clock   151

Fig. 5.24   Photograph of the two stage 6.8 GHZ amplifier prototype.

The completed module is also shown in Fig. 5.24. When the amplifier is first powered up, a resonant frequency of 9 GHz is observed on the spectrum analyser (SA), even without the test signal at the input. However, if the cover is removed, nothing is observed on the SA. It is identified as the cavity resonance that is introduced by the amplifier. The resonant frequency is detected only when the amplifier compartment is enclosed. Furthermore, the resonant frequency can be observed on the SA, when only the amplifier is powered. The effect of cavity resonance is mainly due to the module height and the housing structure. The standing wave has the characteristic, such that the E and H fields are 90° out of phase with each other. The impedance, therefore, fluctuates wildly across the cavity, causing unknown effects on the circuitry, including the introduction of instability to the active devices [170].

It is difficult to simulate the impact of the cavity, because all the properties are not defined accurately, especially the active components. A more practical approach is to examine and identify the cavity resonance on the actual hardware. Relocating a particular circuit element to a different position in the cavity can often fix the problem, but it involves an investment in engineering redesign time and possible manufacturing delays. Hence, a few simpler approaches are considered to solve this problem. Pasting of microwave absorber on the module cover is the quickest and the easiest solution to get rid of spurious cavity resonance. The absorber placement is determined, based on the cut-and-try trial and error method. This solution might incur additional cost to purchase space grade absorber or to conduct qualifying tests for the absorber and adhesive material. In addition, the output power is a few dB lower due to the wave absorption.

5.6.3. Experimental results

In this section, the measured performance of the two-stage amplifier is presented. The total DC power consumption of this amplifier is 123 mW at $V_{ds} = 3.5$ V and $V_{gs} = -3.5$ V, which includes $I_{ds} = 35$ mA in the two stage. The S-parameters of the measurement are performed by the PNA-X Agilent N5244A (10 MHz–42.5 GHz) Network Analyser. A gain of 21 dB is obtained at 6.834 GHz. The input return loss (S_{11}) is lower than -15 dB, which reveals the minimum value of -18 dB at 6.834 GHz. A -16 dB output return loss (S_{22}) is achieved at 6.8 GHz. A small signal gain (S_{21}) is simulated and we observe that the measured gain is lower than the simulated one, and it has similar tendency for all observed frequency ranges. It shows that the measured gain at frequency of 6.8 GHz is 21.35 dB, whilst the simulated one is 26.67 dB. The discrepancy may be due to the different parameters of alumina dielectric substrate, and dielectric loss, used for realizing the prototype. These may have higher values than that used in the simulation. When the value of dielectric loss is higher, some amount of energy from the input port that should actually be transmitted to the output port, is absorbed by the dielectric substrate. This results in smaller value of the measured gain. The discrepancy in the results in the experimental characterization is also observed, for the values of return loss at the input and output ports. In general, both return loss values are shifted to higher frequency range affecting the return loss values at frequency of 6.8 GHz. The measured input and output return loss at frequency of 6.8 GHz are -18 dB and -16 dB whilst the simulation values are -27 dB and -26 dB respectively. The difference in the measured results of return loss is probably, because of the different value of relative permittivity of dielectric substrate. For the result indicated

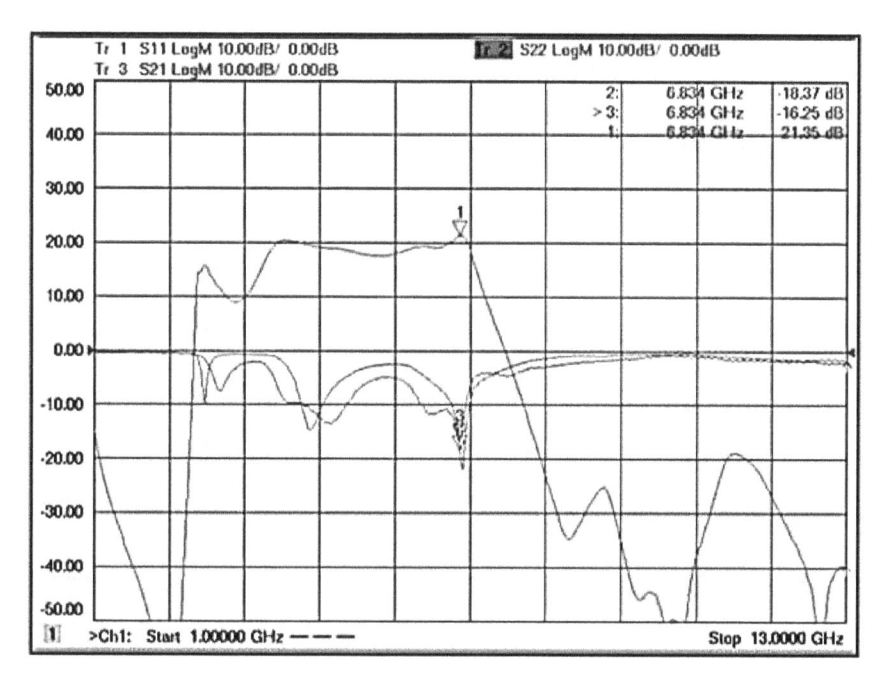

Fig. 5.25 Measured S-parameters (magnitude) of the two stage amplifier CFY-67 against frequency at a drain bias voltage Vds =3.5 V.

in Fig. 5.25, the actual value of relative permittivity seems to be lower than that of the simulation. If the actual relative permittivity is lower, the impedance of matching networks, (i.e., micro-strip lines) move to higher side, resulting in a decrease in the return loss. The small signal gain at 6.8 GHz for the simulated and the measured results are 26 dB and 21 dB, respectively. It is seen from the input and output return loss, an increase in the gain and bandwidth can be expected from a redesign of the matching networks.

Finally, the measured and simulated RF performance results of 2-stage 6.8 GHz amplifier are summarized in Table 5.4.

The design of a 2-stage amplifier for X-band, demonstrates the inter-stage-conjugate-matching (ICM), which is realized on an alumina dielectric substrate. It is shown that utilization of ICM technique for developing the 2-stage amplifier may simplify the calculation of input R/L and output R/L for the matching impedance

Table 5.4 Summary of simulated and measured RF characteristics in two-stage amplifier.

Two Stage Amplifier	Simulated	Measured
Bandwidth	4 GHz–7 GHz	4.2 GHz–7.5 GHz
S21	26.66	21.35 dB
S11	< −27 dB	< −18 dB
S22	< −27 dB	< −16 dB

network. From characterization results, it is observed that the realized 2-stage amplifier prototype has the gain of 21.32 dB at 6.8 GHz, with the power values of input and output of −18 dB and −16 dB, respectively.

5.7. Design of medium and high power 90 MHz and 360 MHz amplifier

5.7.1. *Design methodology for amplifier*

During the sub-module simulation, the passive circuit elements including bias tee, input matching and output matching circuit are optimized and a compromise is achieved between the gain and the stability. The stability factor is given a higher weight to ensure good circuit performance over the gain in the operating frequencies. The simulated Rollett stability factor K for the amplifier is greater than 1 for all frequencies between DC and 1 GHz. After designing the schematic, the matching and bias networks are separately modelled, using an electromagnetic (EM) simulator in ADS.

5.7.2. *Simulation and experimental results of 90 MHz amplifier*

The simulated stability factor, gain and input-output return loss, with a single transistor (BFY-193), are analysed. Since, the circuit is measured in 50 Ω environment, the magnitude of S_{21} directly provides the gain value in the small signal region which corresponds to the linear gain of the large signal measurement. It exceeds 20 dB in the frequency range from 80 to 100 MHz. The measured amplifier

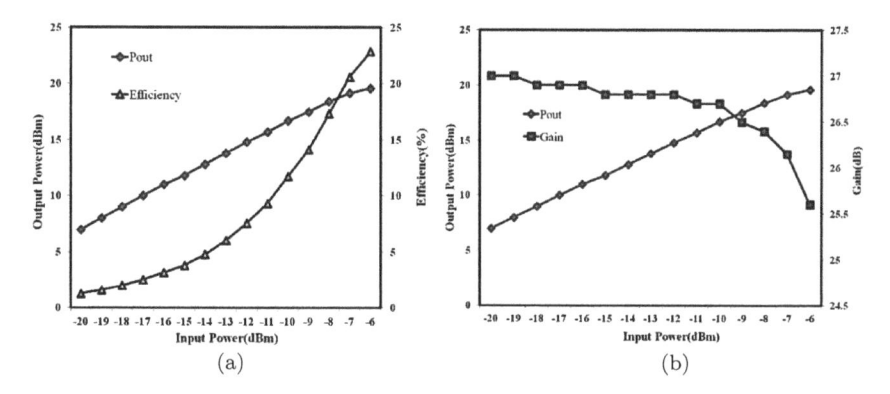

Fig. 5.26 Measured output powers, efficiencies and gain of 90 MHz amplifier (a) Input power vs output power and efficiency plot (b) Input power vs gain plot.

output power, efficiency and gain under variable input conditions are shown in Fig. 5.26.

5.7.3. *Simulation and experimental results of 360 MHz amplifier*

The simulated S-parameters for the 360 MHz amplifier are obtained. The data are collected under DC bias for a 9 V collector bias and 50 mA of quiescent current. The small signal gain is 21 dB in a 300–400 MHz band and the input and output return losses are less than −30 dB and −6 dB respectively. The applied RF input signal is CW, the collector bias voltage is fixed for a Ic = 50mA, amplification of the 360 MHz signal provides the necessary power of 20 to 22 dBm to the SRD's input. This amplification is probably the most sensitive section of the RF synthesizer. In the linear operation regime, the chosen amplifier's gain is slightly more than 20 dB. The 1 dB compression point occurs at an input value of 2 dB, where the amplifier output is approximately 22 dBm. For a 2 dBm input signal, the measured output power, typically exceeds 21 dBm at 360 MHz, which is greater than 20 dBm, essential to drive the SRD. The measured amplifier gain and output power, under variable input conditions, are shown in Fig. 5.27.

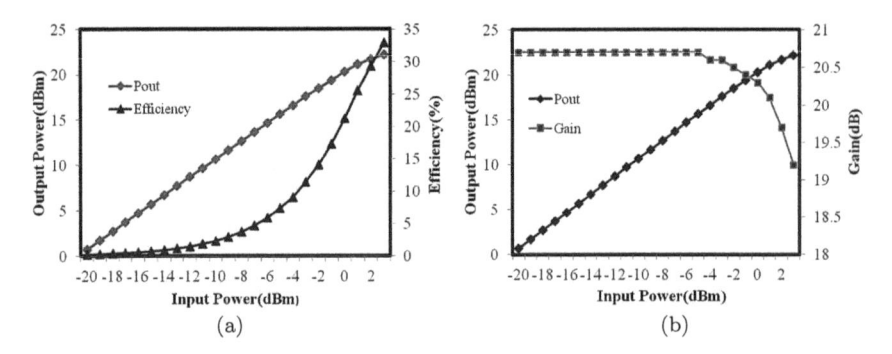

156   Rubidium Atomic Clock: The Workhorse of Satellite Navigation

Fig. 5.27 Measured output powers, efficiencies and gain of 360 MHz amplifier (a) Input power vs output power and efficiency plot (b) Input power vs gain plot.

Table 5.5   The measured results of RF Synthesizer.

Sr.No.	Parameter	Unit	Measured Results
1	O/P Frequency	GHz	6.834687500
2	O/P Power	dBm	+4
3	Harmonics	dBc	−30
4	Spurious	dBc	−40
5	SSB Phase Noise @ offset From centre frequency		
	1 Hz		−49
	10 Hz		−62
	100 Hz		−76
	1 KHz	dBc	−81
	10 KHz		−96
	100 KHz		−103

The compression characteristics of the amplifier shows that its linear gain is 20 dB. It broadly agrees with the simulated results. The linear decrease in the gain, before the onset of compression, is caused by device heating due to the amplifier power dissipation. Hence, the channel temperature also increases linearly as a function of the input power. The measured results of the RF systems are given in Table 5.5.

5.8. Causes of frequency offset

5.8.1. Spurious RF harmonics

In the atomic clock, the microwave spurious harmonics create spurious sidebands. As these are not resonant with the atoms, these lead to a decrease in the power. For narrow band microwave resonance, the spurious sideband signals, that happen to overlap with Zeeman or other atomic resonance, can cause frequency pulling due to unwanted transitions. The spurious RF spectral components can pull the locked frequency, causing a shift in the microwave excitation. Thus, resulting in significant frequency shifts in high performance frequency standards [171–172].

The amount of pulling depends on the relative spurious level, its asymmetry and its separation from the carrier. The relative change in the frequency due to a spurious component is given by

$$\frac{\Delta f}{f} \cong \frac{1}{2} \frac{\gamma^2 B^2}{4\pi^2 f_0 (f_0 - f_1)},$$

where γ the gyromagnetic ratio and B is is the spurious microwave magnetic induction at frequency f_1 [173–175].

A SSB component at 10 MHz output, where spurious level is down by 85 dB below the carrier at 34 kHz offset. When the same carrier is multiplied by 684 to get the Rb resonant frequency, then a relative level of -66 dBc is obtained, that causes a frequency offset of 8.5×10^{-9}. Such a pulling effect could be caused by slightly asymmetrical sidebands and the ripples in the switching power supply. Figure 5.28 shows a single sideband spur, which is 60 dB below the carrier for the Rb frequency standard, operated at a 2.5 dBm optimum microwave power.

5.8.2. PM modulation distortion

The modulation distortion is a primary cause of frequency offsets and instability in passive atomic frequency standards. The low frequency phase modulation (PM or FM) is applied to the Physics package's RF excitation to produce a discriminator signal. This error signal is synchronously detected, and used for generating a control voltage to lock

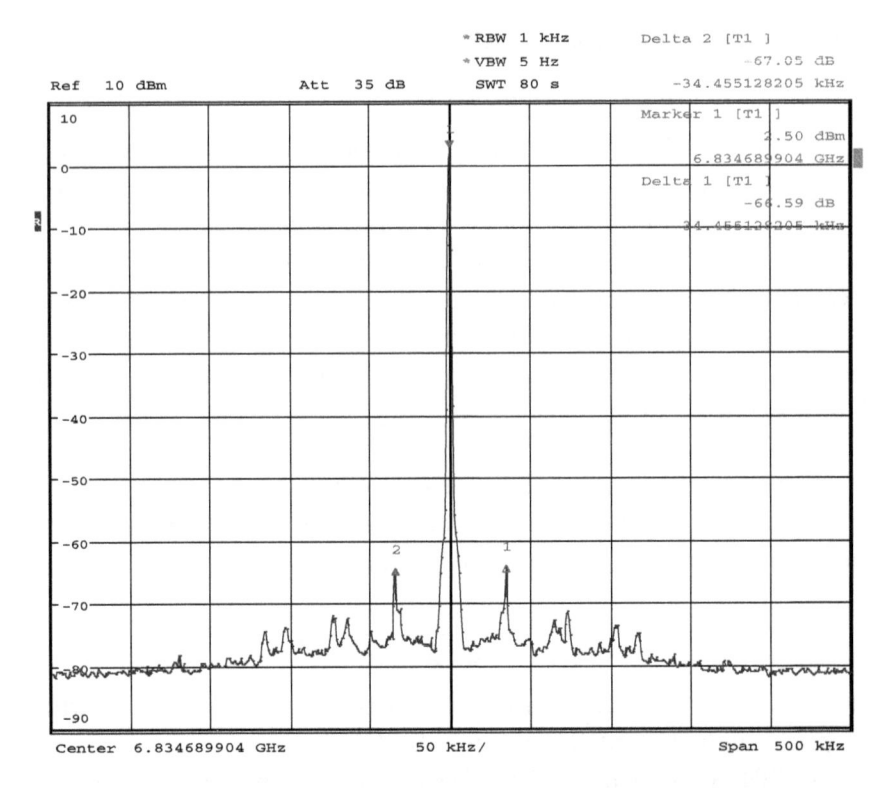

Fig. 5.28 Power spectrum of 6.834 GHz output of Rb synthesizer.

a crystal oscillator to the atomic transition. Even-order modulation distortion causes a frequency offset due to an inter-modulation effect [176]; any change in this offset causes a frequency change. The second harmonic distortion sidebands are produced on the microwave at $\pm 2f_{\text{mod}}$, which interact with the normal modulating signal and generate a spurious fundamental signal, that causes a frequency offset given by

$$\frac{\Delta f}{f} = \frac{\delta_2}{2Q_l},$$

where δ_2 the relative amount of second harmonic distortion and Q_l is the Rb transition line Q [177]. Figure 5.29 shows a -33 dB second harmonic distortion level in a RFS with a 600 Hz line width, this produces a fractional frequency offset of 2.2×10^{-9}. A 5% change

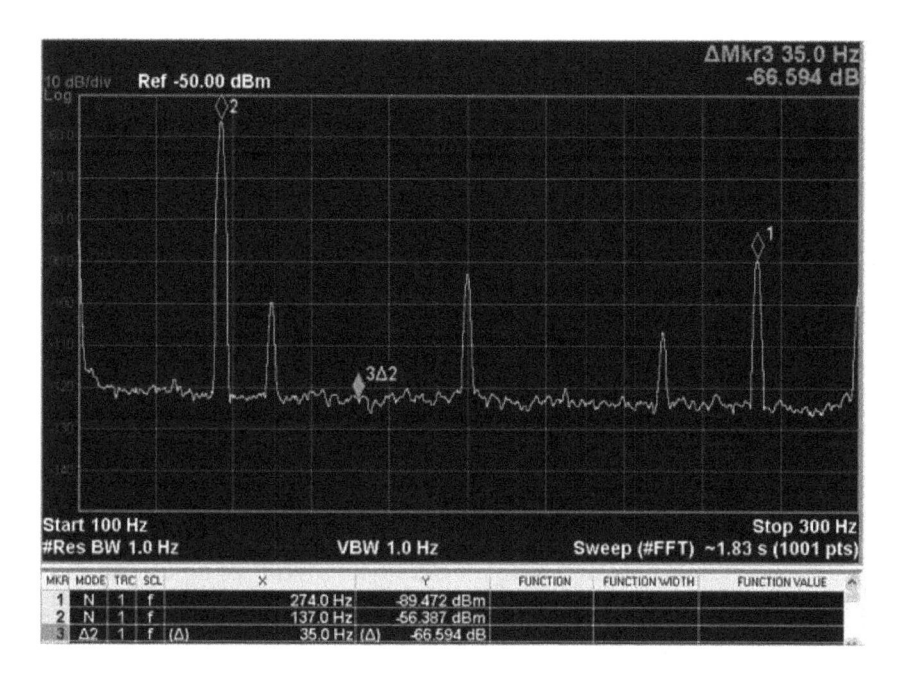

Fig. 5.29   Power spectrum of second harmonic distortion level.

in the amount of distortion, due to an environmental effect, would result in a frequency change of 1×10^{-10}.

The modulation distortion is introduced in several ways: distortion on the modulation signal, distortion in the phase modulator, distortion introduced by asymmetrical RF selectivity and AM-to-PM conversion in the multiplier chain. The modulation signal is made very pure, free from even-order distortion, by generating it from a precise square-wave, followed by passive filtration and integration. The low-distortion phase modulation is possible with a hyper-abrupt tuning varactor, in an all-pass network. The phase modulation is done at a relatively low RF in the frequency multiplier chain, where the required deviation is low. An active phase modulator, such as a phase-lock loop (PLL), can introduce distortion because of coherent ripple in the modulation transfer function. A passive network is generally a better option. There are many subtle modulation distortion effects, which can occur in a RF multiplier chain. Each stage of a harmonic multiplier enhances the PM index and can suppress AM by using amplitude limiter. The AM-to-PM and PM-to-AM conversion can cause frequency sensitivity to RF stage tuning and its level. The PLL multipliers can have problems due to finite loop bandwidth and phase detector distortion. The step recovery diode (SRD) multipliers exhibit sensitivity to the input power and bias conditions. The first stage in a multiplier chain is usually the most critical, since that is where the AM and PM indices are the closest. The spurious components are closest to the carrier. The inter-stage selectivity is critical in a harmonic multiplier chain; it is vital to avoid spectral asymmetry caused by complex mixing between sub-harmonic components. The output of each stage must be well-filtered before driving the next stage, and selected networks must be symmetrical and stable against temperature and drift. It is especially important to have a pure drive signal for the final SRD multiplier. A direct multiplier chain is preferable to one which uses the frequency mixing, to avoid asymmetrical microwave spectral components. The modulation of the VCXO by second harmonic ripple from the servo amplifier has the same effect as even order modulation distortion, producing a frequency offset that depends on environmental conditions.

5.8.3. Amplitude modulation distortion

The amplitude modulation on the microwave excitation is another form of the modulation distortion, that can cause frequency offset and instability. The frequency offset caused by AM at the servo modulation frequency is given by

$$\frac{\Delta f}{f} = \frac{\alpha_1}{2Q_l},$$

where α_1 is the relative amount of AM [177]. In the second harmonic PM distortion, a -70 dB AM level with a Rb line Q_l of 23×10^6, results in a frequency offset of 7×10^{-12}.

5.8.4. Sub-harmonics distortion

The sub-harmonics are a particularly bothersome spectral component in the input signal to the SRD multiplier. The sub-harmonic spectral components introduce time jitter between the impulses that generate the microwave energy, and can change the average RF power as the spectrum changes versus temperature or some other environmental conditions. The period of the Rb microwave excitation is about 150 ps, so the time jitter of the SRD multiplier input waveform of the order of 4 ps (only about **0.5°** at 360 MHz) can have a significant effect on its amplitude. The change in the microwave power in dB is given by,

$$\Delta P = 20 \log 10 \left[\left| \cos(2.10^{\frac{S}{20}}.N) \right| \right],$$

where S is the sub-harmonic level of the drive to the SRD multiplier in dBc and N is the multiplication factor. The expression describes an interference pattern that has nulls for certain conditions of the cyclic phase jitter. The argument of the cosine function is in radians. Even a relatively "clean" SRD multiplier drive spectrum can introduce significant frequency offsets. For a typical multiplication factor of 19, the sub- harmonics must be suppressed by at least 25 dB, shown in Fig. 5.30 to insure that the microwave power is stable withing ±0.5 dB against small changes in sub-harmonics on the multiplier drive.

5.9. Phase noise measurement

The RF multiplier or synthesizer's phase noise sets the theoretical stability limit of an atomic clock. The phase noise at higher order harmonics of the modulation frequency f_m, results in a limit to the clock frequency stability [178], given by:

$$\sigma_Y(\tau) = \sqrt{\sum_{n=1}^{\alpha} C_{2n}.S_\phi(2nf_m)}.\tau^{-1/2}, \qquad (5.7)$$

where n is the harmonic number, S_φ is the phase noise measured at even harmonics of the modulation frequency, and C_{2n} is the

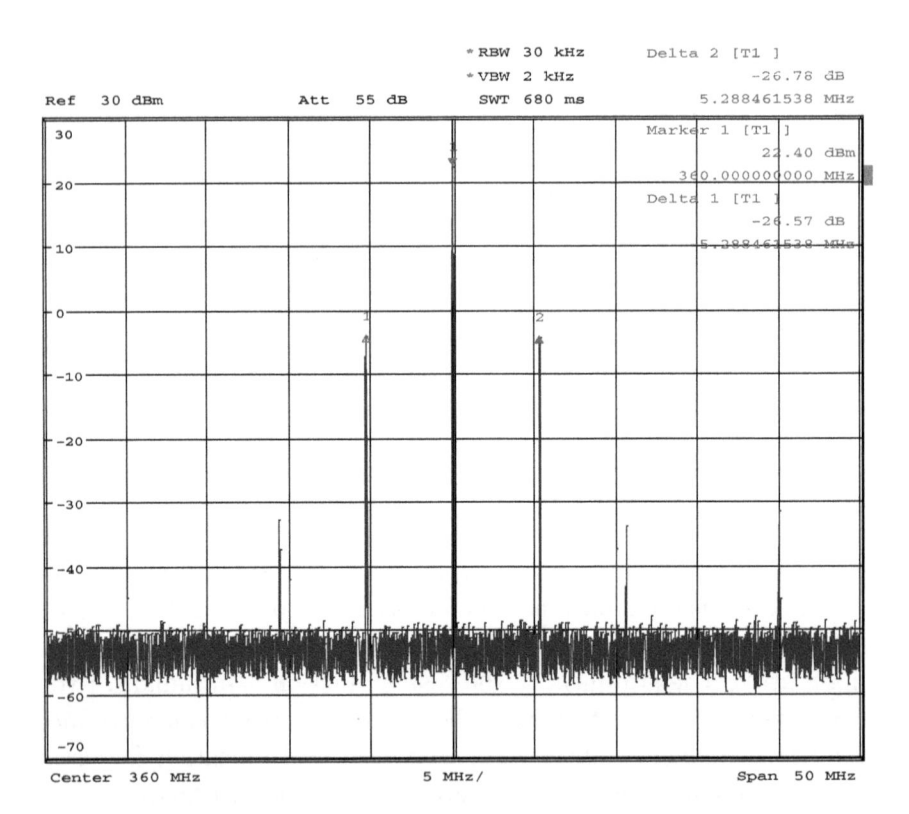

162 Rubidium Atomic Clock: The Workhorse of Satellite Navigation

Fig. 5.30 Power spectrum of sub harmonic signal to the SRD.

coefficient of sensitivity given by:

$$C_{2n} = \frac{2n}{(2n-1)(2n+1)} \frac{f_m}{\nu_0}, \qquad (5.8)$$

where ν_0 is the multiplier or synthesizer output frequency, $f_m = 137$ Hz in the present case. The measured phase-noise of -130 dBc/Hz at 274 Hz is shown in Fig. 5.31 we obtain a short-term stability limit, due to the Dick effect [17], $\sigma(\tau) = 7.7 \times 10^{-12} \tau^{-1/2}$.

This stability limit is well within the typical short-term stability specification of $5 \times 10^{-12-} \tau^{-1/2}$ for the space atomic clocks. The phase noise of the optimized synthesizer is measured using a phase noise measurement system. Figure 5.32 reports the measured single-sideband phase-noise, for the RF synthesizer at 6.8 GHz, reaching ≈ -87 dBc/Hz in the band of interest from 100 Hz to 1 kHz (upper,

RF Synthesizer for Rb Atomic Clock 163

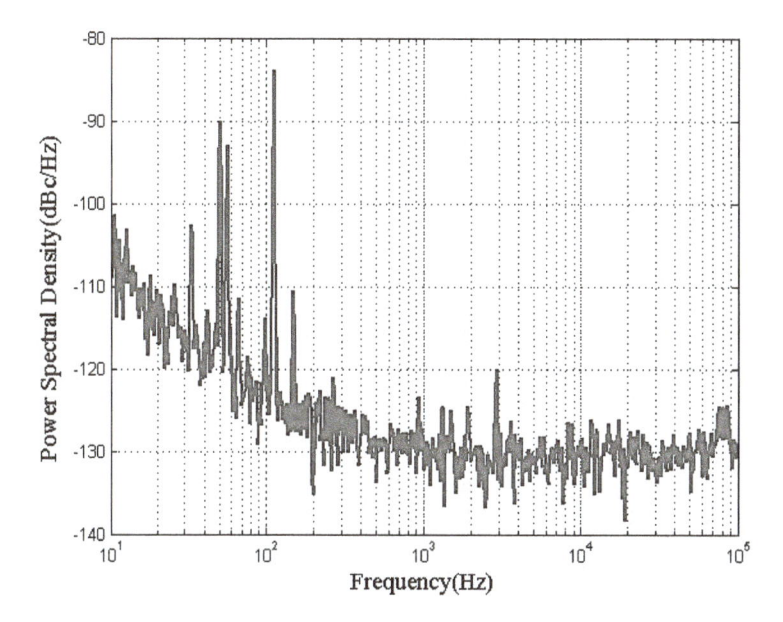

Fig. 5.31 Measured output phase noise spectrum of the 10 MHz OCXO.

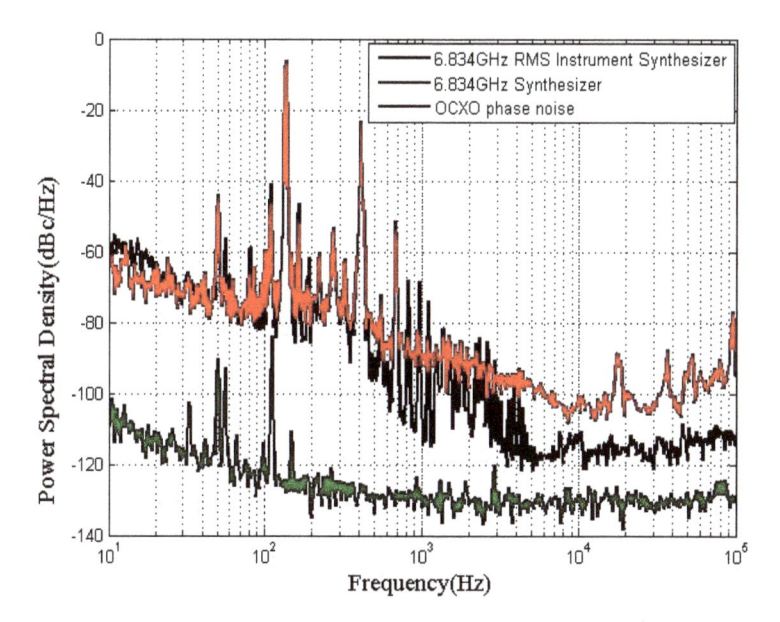

Fig. 5.32 Measured output phase noise spectrum of R&S instrument (black curve) and clock synthesizer (red curve) at 6.834 GHz. The green plot shows the phase noise spectrum of OCXO at 10 MHz.

red trace). This phase-noise level is only slightly higher, 10 dB to a maximum 15 dB, than the phase-noise of the 6.834 GHz RMS instrument synthesizer (black trace).

The 10 MHz thermally-compensated oven controlled quartz oscillator (OCXO) phase noise transferred to 6.8 GHz carrier (lower, green trace). This result proves, the good phase-noise performance of the synthesizer, and especially, an improvement compared to the initial synthesizer with raditek OCXO, which showed a phase noise of −67 dBc/Hz @ 100 Hz offset and −77 dBc/Hz @ 1 kHz offset from the carrier.

Summary

A low noise microwave synthesizer development for the Rb atom clock is discussed in this chapter. The characterization and the optimization of the synthesizer results in a low flicker phase noise of −77 dBc/Hz at offset frequency of 274 Hz. This corresponds to the frequency stability of 2.5×10^{-12} at 1s which better than the level estimated for the clock. The clock frequency stability requirement is 5×10^{-12} or better at 1s, which is consistent with the stability of the system.

Chapter 6

Design Simulation and Development of Microwave Cavity

The microwave cavity size is an important factor in the overall dimensions of the Rb atomic clock. It is a challenge to design and develop the microwave cavity for high Q, small dimensions and suppression of spurious cavity modes. In many of the Rb atomic clocks, TE_{111} cavity mode is used. We will mainly discuss this mode with practical details to help those, who want to manufacture the Rb atomic clocks. The other types of microwave cavity are being conceptualized but they are in experimental stages. These cavities are discussed briefly in Chapter 10.

6.1. Introduction

A microwave cavity is an essential part of a Physics package of any passive Rb frequency standard, and its magnetic field configuration plays very important role in performance of the clock. The unloaded Q is one of the main issues in designing the microwave resonator. Its characteristics naturally depend on the electromagnetic field distribution. The passive Rb atomic clocks are based on the double-resonance spectroscopy. That requires two electromagnetic fields: an optical field at $\nu_{opt} = 384.23$ THz, (Rb D2-line) for creating population inversion by the optical pumping, and the microwave field at ν_{Rb} (unperturbed value equals to 6 834 682 610.90 429(9) Hz) to drive the g.s hyperfine clock transition. Traditionally, the microwave cavity for the passive Rb frequency standard is a cylindrical resonator that operates either in the TE_{111} mode or TE_{011} mode. TE_{011} mode produces a fairly uniform magnetic field near the centre, along the cylindrical axis. However, in order to resonate at the 6.834 GHz, the

Rb hyperfine frequency, it must have an inner diameter larger than 5 cm. This represents severe constraint in minimizing the size of Rb frequency standard. A cylindrical cavity operating in the TE_{111} mode offers an alternative. The magnetic field configuration in TE_{111} mode cavity is the maximum at the cavity walls and tapering to zero at the centre of the cavity [179]. The TE_{111} resonant mode is generally used, as for a given frequency, length to diameter ratio >2. Besides, the cylindrical TE_{111} cavity diameter is about 1/2 that of the TE_{011} cavity. There are several microwave cavities considered for the Rb atomic frequency standard in order to reduce the cavity size. Earlier, cavity designs have used a rectangular cavity. One such design introduced by H.E William *et al.* [179] is a rectangular TE_{011} cavity, partially loaded with dielectric, resulting in a considerable size reduction. In other configuration described in the US Patent N0.4, 349,798 [180], a rectangular main cavity operating in TE_{021} mode, with two secondary cavities on opposite sides of the main cavity, was used to produce lumped resonant loading. Yet another small size magnetron-type cavity operating in a pseudo cylindrical TE_{011} mode, was described by Schweda *et al.* [181]. However, the magnetron cavity has a more complicated structure, and it is difficult to fabricate [182]. A TE_{101} rectangular cavity partially loaded with a dielectric slab was described in US. Pat. N0.4, 495,478 [183]. Other microwave cavities were also designed, which used lumped LC resonators and produced longitudinal microwave magnetic fields. Its example was a helical cavity resonator, used for Rb frequency standard, US. Pat. Nos. 4, 947,137 [184]. Other resonators of this type were "split ring" (or "loop gap" or "slotted tube") resonators [185], and were described by W. N. Hardy and L. A. Whitehead and U.S. Pat. Nos.4, 446,429 and 4,633,180. A new type of the slotted tube resonator has recently been used in a Rb frequency standard [186]. A novel design of a rectangular-cylindrical microwave cavity, that takes into account the presence of the glass cell and use SRD multiplier circuit is discussed by E. Eltsufin *et al.* [187].

Theoretically, a simple cavity completely, enclosed by metallic walls can oscillate in any one of the infinite number of field configurations. The free oscillations are characterized by an infinite number of resonant frequencies, corresponding to the specific field patterns of modes of oscillation. The cavity dimensions and electromagnetic properties of cavity loads [188] determine the resonance frequency of each cavity mode. Although the reflection and transmission characteristic plots give the number of modes into the cavity, but they do not indicate exactly, which modes are present. This situation is even more challenging when many modes are present. Generally, the decision to identify resonant mode, strongly depends on the dimension of monitoring probe and its position. Besides, the relative amplitude resonances are variable for different loop locations and dimensions. The feed loop also tends to shift the modes and sometimes splits degenerate modes [188]. The work reported in this chapter, first describes the conducting and dielectric loss of the metal cavity for calculating the loss tangent measurement. The quality factor is calculated from the loss tangent of the dielectric sample, as well as by conducting loss of the metal cavity. The radiation loss in the cavity can be completely eliminated by placing the dielectric material inside the cavity. When a high dielectric material is put in the centre of a metal cavity, the electromagnetic field gathers around the dielectric. As a result, the resonant frequency becomes lower, the conductor loss on the metal surfaces becomes less that appears to be due to the loss in the dielectric [189]. This is also verified and compared with HFFS simulation for the cavity. In this chapter, the reflection and transmission characteristics, based on the loop insertion into cylindrical metallic cavity, is simulated and experimentally verified for both TE_{111} and TE_{011} modes. The chosen position and dimensions of loops, allow comparison of both reflection and transmission characteristics. The influence of loop size on the resonant frequency of TE_{111} and TE_{011} modes is taken into account. This theoretical and simulated field model clearly describes the field distribution inside the resonator. Finally, it is shown that such a design makes a TE_{111} cavity very small, while maintaining high magnetic field uniformity. It is adopted for space qualified Rb atomic clock.

6.2. Theoretical analysis of cavity Q

Various theoretical analyses on this cylindrical dielectric resonator have been published [188, 190–194]. The structure is also applied to loss tangent measurement by simply neglecting the conducting loss of the metal cavity [195–196]. Many of these theories involve quite complex analysis. To simplify and understand the field analysis of the cylindrical dielectric resonator, this study adopts a simple field model. This electromagnetic field of the TE_{111} mode for the circular cylindrical resonator can be expressed as [197–199].

$$H_z = H_0 J_1\left(\frac{p'_{11}\rho}{a}\right)\cos\phi \sin\frac{l\pi z}{d}, \tag{6.1a}$$

$$H_\rho = \frac{\beta a H_0}{p'_{11}} J'_1\left(\frac{p'_{11}\rho}{a}\right)\cos\phi \cos\frac{l\pi z}{d}, \tag{6.1b}$$

$$H_\varphi = \frac{-\beta a^2 H_0}{(p'_{11})^2 \rho} J_1\left(\frac{p'_{11}\rho}{a}\right)\sin\phi \cos\frac{l\pi z}{D}, \tag{6.1c}$$

$$E_\phi = \frac{jk\eta a H_0}{p'_{11}} J'_1\left(\frac{p'_{11}\rho}{a}\right)\cos\phi \sin\frac{l\pi z}{d}, \tag{6.1d}$$

$$E_\rho = \frac{jk\eta a^2 H_0}{(p'_{11})^2 \rho} J_1\left(\frac{p'_{11}\rho}{a}\right)\sin\phi \sin\frac{l\pi z}{d}, \tag{6.1e}$$

$$E_z = 0 \tag{6.1f}$$

where a is the radius of the cavity, d is the cavity height/length, $k = \omega\sqrt{\mu\varepsilon}$, and $\mu\varepsilon$ are the permeability and permittivity of the material filling the cavity, $\eta = \sqrt{\mu/\varepsilon}$ represents the intrinsic impedance of the material filling and $J_1(p'\rho/a)$ denotes the Bessel functions of order one. The unloaded quality factor of the cavity is expressed as [191],

$$Q = \omega\frac{W_m + W_e}{P_{Dc} + P}, \tag{6.2}$$

where W_m and W_e denote time-averaged stored magnetic and electric field energy respectively, P_d and P_c are the power dissipated in the dielectric and conducting wall, respectively. At resonance $W_m = W_e$ and we have,

$$Q = \omega \frac{2W_m}{P_{loss}}, \quad (6.3)$$

where $P_{loss} = P_d + P_c$ is the total dissipated power. Since the time-averaged stored electric and magnetic energy are equal, the total stored energy is,

$$W = 2W_e = \frac{\varepsilon}{2} \int_{z=0}^{d} \int_{\phi=0}^{2\pi} \int_{\rho=0}^{a} (|E_\rho|^2 + |E_\phi|^2) \rho d\rho d\phi dz$$

$$= \frac{\varepsilon k^2 \eta^2 a^2 \pi d H_0^2}{4(p'_{11})^2} \int_{\rho=0}^{a} \left[J_1'^2\left(\frac{p'_{11}\rho}{a}\right) + \left(\frac{a}{p'_{11}\rho}\right)^2 J_1^2\left(\frac{p'_{11}\rho}{a}\right) \right] \rho d\rho$$

$$= \frac{\varepsilon k^2 \eta^2 a^4 H_0^2 \pi d}{8(p'_{11})^2} \left[1 - \left(\frac{1}{p'_{11}}\right)^2 \right] J_1^2(p'_{11}), \quad (6.4a)$$

and the power loss in the conducting wall is,

$$P_c = \frac{R_S}{2} \int_S |\bar{H}_{\tan}|^2 ds$$

$$= \frac{R_S}{2} \left\{ \int_{Z=0}^{d} \int_{\phi=0}^{2\pi} [|H_\phi(\rho=a)|^2 + |H_Z(\rho=a)|^2] a d\phi dz \right.$$

$$\left. + 2 \int_{\phi=0}^{2\pi} \int_{\rho=0}^{a} [|H_\rho(z=0)|^2 + |H_\phi(z=0)|^2] \rho d\rho d\phi \right\}$$

$$= \frac{R_S}{2} \pi H_0^2 J_1^2(p_{11}^1)$$

$$\times \left\{ \frac{da}{2} \left[1 + \left(\frac{\beta a}{(p'_{11})^2}\right)^2 \right] + \left(\frac{\beta a^2}{p'_{11}}\right)^2 \left(1 - \frac{1}{(p'_{11})^2}\right) \right\}$$

$$(6.4b)$$

Then from Eq. (6.3), the Q of the cavity with imperfectly conducting walls but lossless dielectric is calculated by the equation,

$$Q_c = \frac{\omega W}{P_c} = \frac{(ka)^3 \eta a d}{4(p'_{11})^2 R_S} \frac{1 - \left(\frac{1}{p'_{11}}\right)^2}{\left\{\frac{ad}{2}\left[1 + \left(\frac{\beta a}{(p'_{11})^2}\right)^2\right] + \left(\frac{\beta a^2}{p'_{11}}\right)^2 \left(1 - \frac{1}{(p'_{11})^2}\right)\right\}} \quad (6.4c)$$

To compute the Q due to the dielectric loss, power dissipation in the dielectric is calculated by,

$$P_d = \frac{1}{2} \int_V J.E^* dv = \frac{\omega \varepsilon''}{2} \int_V [|E_\rho|^2 + |E_\phi|^2] dv$$

$$= \frac{\omega \varepsilon'' k^2 \eta^2 a^4 H_0^2}{8(p'_{11})^2} \int_{\rho=0}^{a} \left[\left(\frac{a}{p'_{11}\rho}\right)^2 J_1^2\left(\frac{p'_{11}\rho}{a}\right) + J_1'^2\left(\frac{p'_{11}\rho}{a}\right)\right] \rho \, d\rho$$

$$= \frac{\omega \varepsilon'' k^2 \eta^2 a^4 H_0^2}{8(p'_{11})^2} \left[1 - \left(\frac{1}{p'_{11}}\right)^2\right] J_1^2(p'_{11}) \quad (6.4d)$$

Then from Eqs. (6.3) and (6.4d), Q is calculated,

$$Q_d = \frac{\omega W}{P_d} = \frac{\varepsilon'}{\varepsilon''} = \frac{1}{\tan \delta}, \quad (6.5a)$$

where $\varepsilon = \varepsilon' - j\varepsilon'' = \varepsilon_r \varepsilon_0 (1 - j \tan \delta)$, and $\tan \delta$ is the loss tangent of the material. So from Eq. (6.2) total Q of the cavity is,

$$Q = \left(\frac{1}{Q_C} + \frac{1}{Q_d}\right)^{-1}. \quad (6.5b)$$

There are several approximate calculations, of the resonant frequency of the circular cross-section cavity of TE mode [190–193]. The TE$_{mnp}$ mode with the smallest resonant frequency, is the TE$_{111}$.

Its resonant frequency is given by [192],

$$(f_r)\text{TE}_{111} = \frac{1}{2\pi\sqrt{\mu\varepsilon}}\sqrt{\left(\frac{1.8412}{A}\right)^2 + \left(\frac{\pi}{H}\right)^2}. \quad (6.6)$$

Similarly, the TM_{mnp} mode with the smallest resonant frequency, is the TM_{010}, and its resonant frequency is given by:

$$(f_r)\text{TM}_{010} = \frac{1}{2\pi\sqrt{\mu\varepsilon}}\left(\frac{2.4048}{a}\right). \quad (6.7)$$

In the above equations f_r = resonant frequency, A, a = radius of the cavity, H = height/length of the cavity, μ = permeability of the dielectric filling of the cavity and ε = permittivity of the dielectric filling of the cavity. From Eqs. (6.6) and (6.7), the same frequency is realized when h/a = 2.03 ≈ 2, whereas for h/a < 2.03 the dominant fundamental mode is TM_{010}, and for h/a > 2.03 the dominant mode is the TE_{111} mode. The dimensions of the cavity are fixed such that, for the dominant resonant frequency 6.834 GHz, the cavity mode is TE_{111}. Besides, the dominant mode TE_{111}, others modes coexist in the cavity, some of the lower modes and their resonant frequencies are discussed in Sec. 6.3.

6.3. 3-D cavity simulations in HFFS

We use high frequency field simulation (HFFS) to map the mode distribution and change of electrical and magnetic fields inside the cavity. The development of the field model is also simulated in HFFS for the TE_{111} and TE_{011} modes. The design and simulation comparison for the cavity Q, magnetic field configuration and tuning optimization of resonance frequency take into account the presence of dielectric material such as glass cell, Teflon ring and photodiode detector PCB for both TE_{011} and TE_{111} cavities.

6.3.1. Mode and field simulation for TE_{011} mode

The HFFS simulation is used to produce resonant frequency and eigenvalue graphs for TE_{011} and TE_{111} modes. The first simulation is done for TE_{011} mode cavity to observe the resonant frequency for unloaded cavity, along with magnetic field configuration. Then we thoroughly analyse the change of the resonance frequency and the cavity Q with the addition of dielectric material, in the form of the absorption cell, photodiode PCB and Teflon ring into the cavity. We know that, if H/A > 2.03 (A = radius, H = length), the primary mode in a cylindrical cavity is TE_{111}; otherwise it is TM_{010}. First we simulate the cavity for TE_{011} mode, with the condition H/A or h/a < 2.03. We start with the cavity length of 55.5 mm and cavity diameter of 56 mm to get the resonance frequency of 6.834 GHz by eigen mode simulation. When we insert the absorption cell, the resonant frequency decreases from 6.835 GHz to 5.357 GHz, so to get back the resonance frequency we need to optimize both length and diameter of the cavity simultaneously, to keep the ratio h/a = 2. This is the main drawback for optimization of TE_{011} mode, whereas in the case of TE_{111} mode, we can easily tune it by only changing the length. The cavity length and diameter optimization is done by inserting three different dielectric materials with the permittivity of 2.1(Teflon), 3.8 (absorption glass cell) and 4.4 (FR4 photodiode PCB) FR-4 or FR4. FR4 is a the name of the grade designation assigned to glass-reinforced epoxy laminated printed circuit boards (PCB). Finally, we get the cavity length of 35 mm and diameter of 36 mm for TE_{011} mode, and it is observed by HFFS simulation that its magnetic field configuration is the maximum at the centre with a Q of 240. Using the EM simulation HFFS, cavity modes with dielectric material and the magnetic field distribution of dielectric-loaded cavity are depicted in Figs. 6.1(a) and 6.1(b) respectively.

6.3.2. Mode and field simulation for TE_{111} mode

In case of TE_{111} mode, it is observed from Eq. (6.6). that for the fixed value of the radius and height/length of cavity and the permittivity of the dielectric material; the resonant frequency can be determined. A number of dielectrics with different ε_r values are then tested, with the calculation we find for each of them, the resulting resonant frequencies which are compared with the simulation. In Fig. 6.2, the graph

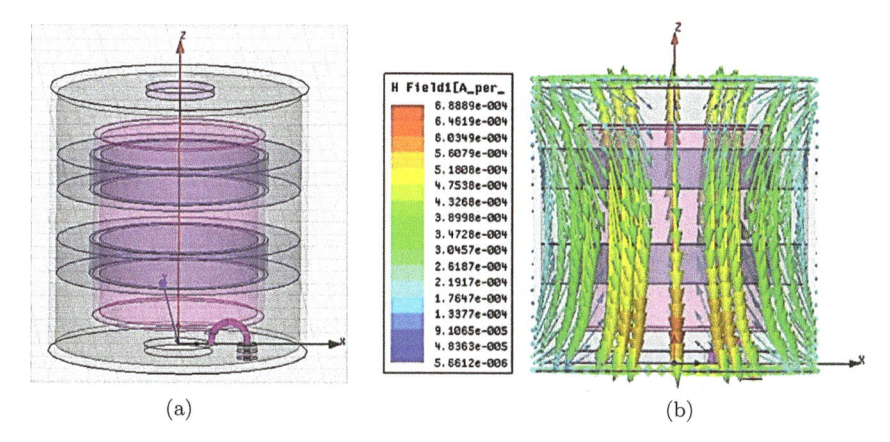

Design Simulation and Development of Microwave Cavity 173

(a) (b)

Fig. 6.1 (a) TE$_{011}$ cavity model with dielectric loading (b) Magnetic field in the meridian plane, mode TE$_{011}$.

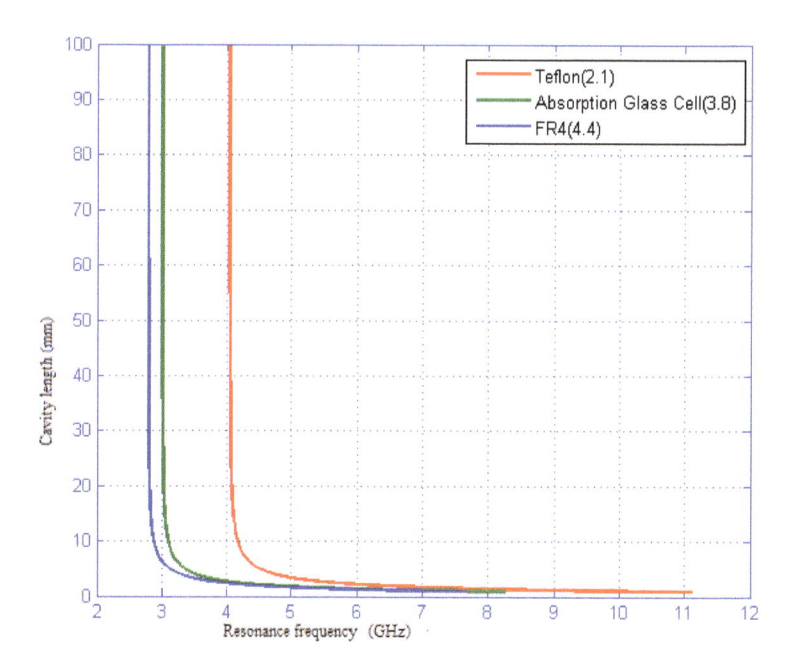

Fig. 6.2 Variation of cavity length versus resonance frequency with three different dielectric constant.

of the cavity length versus the resonant frequency of the cavity is shown for the chosen values of ε_r, 2.1(Teflon), 3.8 (absorption glass cell) and 4.4 (FR4 photodiode PCB).

The first run of simulation in the unloaded cavity gives the fundamental TE_{111} mode with the resonant frequency at 6.834 GHz; other modes that can be observed are TE_{112}, TM_{010}, TM_{011}, TM_{012} and TE_{210}. The next simulation is for the TE_{111} mode resonant frequency variation versus the permittivity of the dielectric insertion. The graph shows that the permittivity of the dielectric inserted in the cavity varies inversely with the resonant frequency. From Fig. 6.2, we observe that a cavity with a volume of $23\,cm^3$ at the resonance frequency of 6834 MHz is expected to be realized by the cavity, loaded with a dielectric material. The dielectric constant of the whole cavity is considered to be less than 2.5. The frequency, where the curve cuts the horizontal or x-axis is the resonant frequency of the TE_{111} mode (the fundamental mode) of the loaded cavity. The analysis of a dielectric loaded cavity is done by inserting three different dielectric materials with permittivity of 2.1(Teflon), 3.8(absorption glass cell) and 4.4(FR4 photodiode PCB) at the middle of the cavity. It is found that due to this dielectric loading, the resonance frequency is reduced. So, to get the resonance back, a length reduction is required and this is done by using a tuning plunger. After optimization, the cavity length is reduced from 44 mm to 31 mm with the radius of the cavity remaining fixed. A diagram of the unloaded and dielectric loaded cavity is shown in Fig. 6.3.

Though the size is reduced, the wall loss becomes larger owing to the strong magnetic field on the conductor wall. To reduce the conductor loss, the magnetic field distribution has to be kept away from the conductor wall and for reducing the dielectric loss, the electric field distribution has to be displaced from the dielectric to low loss region. In practice, the supporting rings of Teflon are inserted between the absorption cell and the cavity wall. Due to these two Teflon rings, the electromagnetic field enters the dielectric material (Teflon ring), and field distribution of the resonance mode is deformed. In general, the dielectric material of high permittivity is chosen, which offers the advantage of reduced conductor

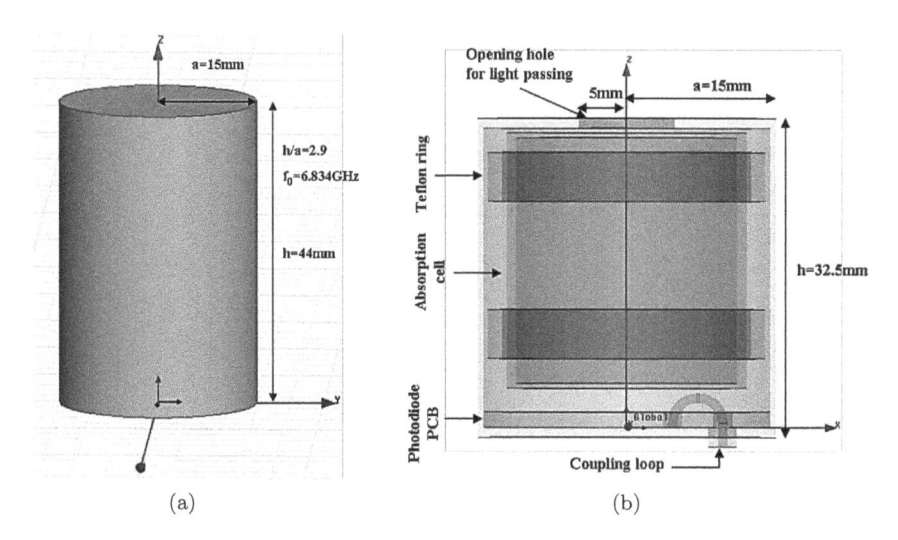

Design Simulation and Development of Microwave Cavity 175

Fig. 6.3 The cavity dimensions (a) without dielectric loading and (b) with dielectric loading. Dimensions not to be scaled.

losses. Because the electromagnetic field is effectively, confined to the dielectric regions. Using the EM simulation HFFS, the electric and magnetic field distribution of dielectric loaded cavity are depicted in Fig. 6.4 for TE_{111} mode.

6.3.3. Cavity tuning and loop coupling

For the practical realization of the optimum transfer of energy, one of the most important issues is the resonant mode distribution inside the metallic cavity. The knowledge of the propagating modes helps in determining the location of the input coupling structures in the field and tuning devices. There are many ways to couple energy into the cavity [176, 179, 185, 186–188, 189]. Here we couple the energy into the cavity via the method of loop coupling [194, 197] as shown in Fig. 6.3. In TE_{111} mode, a large portion of the electric field to propagate along the length of the resonator, causing high currents and therefore, losses in the dielectric material. Silver coatings are used to prevent the radiation losses. Finally, the loaded Q ≈ 200 meets the passive atomic clock requirements [198] and ensures a negligible cavity pulling contribution on the atomic signal [199]. The copper

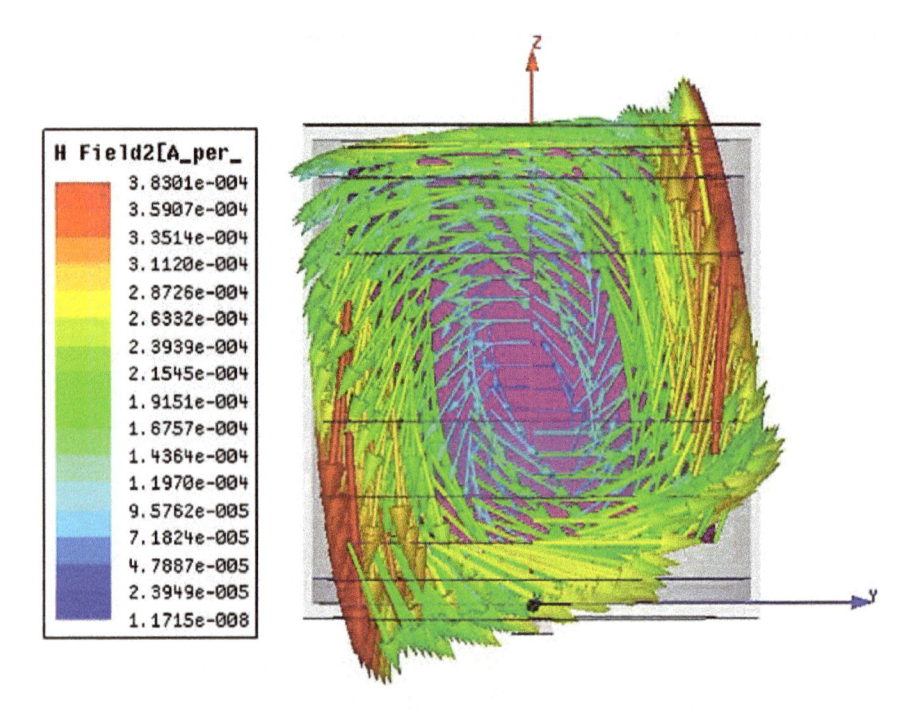

176 *Rubidium Atomic Clock: The Workhorse of Satellite Navigation*

Fig. 6.4 Magnetic field in the meridian plane-mode TE$_{111}$.

wire loop is inserted into the cavity and its free end is connected to the cavity wall. For a resonator operating in TE$_{111}$ mode, we use loop coupling that has some length parallel to the z-axis of the resonator, thus intercepting more of the electric field. It is shown [200] that the resonant frequency might vary with the length of the inserted loop. This leads to the importance of simulating the cavity using different lengths of coupling loop. The simulation helps in choosing the correct loop length and the results are close to the practical value. Five probe lengths are tested in the simulation to give the input return loss (S_{11}dB) level and the resonant frequency for loaded cavity. The results are given in Table 6.1 and plotted in Fig. 6.6.

It is seen from the above results, that the probe length, which gives the least return loss in transmission, is 7 mm. The 8 mm probe shows the best result in term of the desired resonant frequency, being the closest to the atomic resonant frequency 6.834 GHz. As the correct

Table 6.1 Loop length and the corresponding S_{11} and f_0.

Loop Length	S_{11}dB	f_0(GHz)
6 mm	−24.62	6.83405
7 mm	−24.50	6.83539
8 mm	−23.79	6.83452
9 mm	−23.58	6.83501
10 mm	−23.50	6.83514

resonant frequency is more crucial than the losses, the 8-mm probe is used in rest of the simulations. The diameter of the loop is about 2.5 mm.

6.4. Cavity performance and measurement results

Two designs of the cavity are simulated in HFFS, and performance of TE_{111} mode cavity is measured. The TE_{111} mode cavity is made of aluminium and silver-plated internally. The resonant frequency of the cavity is engineered to be slightly higher at the ambient temperature, in order to allow the precise tuning at the desired temperature with a servo-controlled heater. As an illustration, the role of probe length to the level of injected EM field, the experimental results of EM field strength, detected from input return loss (S_{11}) for the dominant TE_{111} mode, are given in Table 6.1. The experimental results confirm that the input return loss (S_{11}) is reduced for the probe length, which is $\lambda/4$ of the resonance frequency. The input return loss S_{11} of the cavity is measured to be −23 dB with 5 dB attenuator pad, that simplifies its matching with the cavity input coupling port. Figure 6.5 shows the realized prototype microwave cavity with a tuning plunger and Teflon ring. Using the output matching scheme with the plunger, for a 6.834 GHz cavity operating in the TE_{111} mode, results are shown in Figs. 6.6–6.8.

The curve in Fig. 6.6 is produced by varying the tuner length connected to the output port until the measured impedance, at the input reference plane is close to Z_0, 50 ohm at $f_0 = 6.834$ GHz. This can be seen on the Smith Chart of Fig. 6.7, where the curve intersects

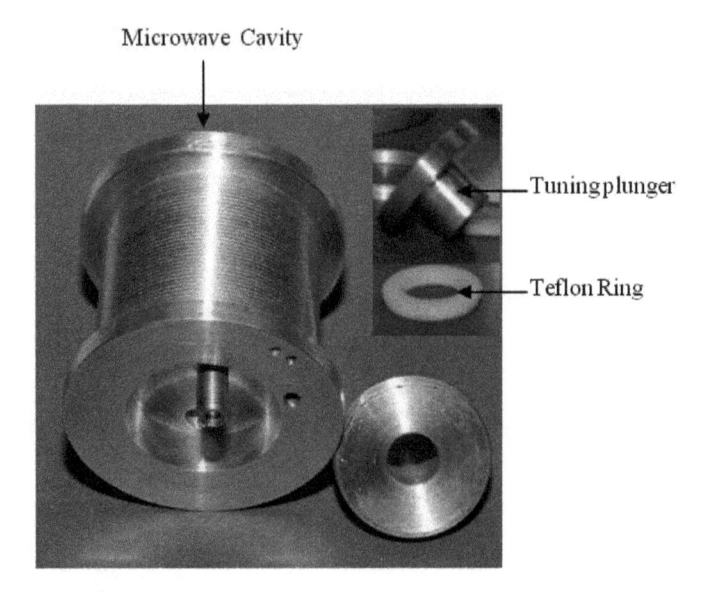

178   Rubidium Atomic Clock: The Workhorse of Satellite Navigation

Fig. 6.5 Realized prototype microwave cavity with tuning plunger and Teflon ring.

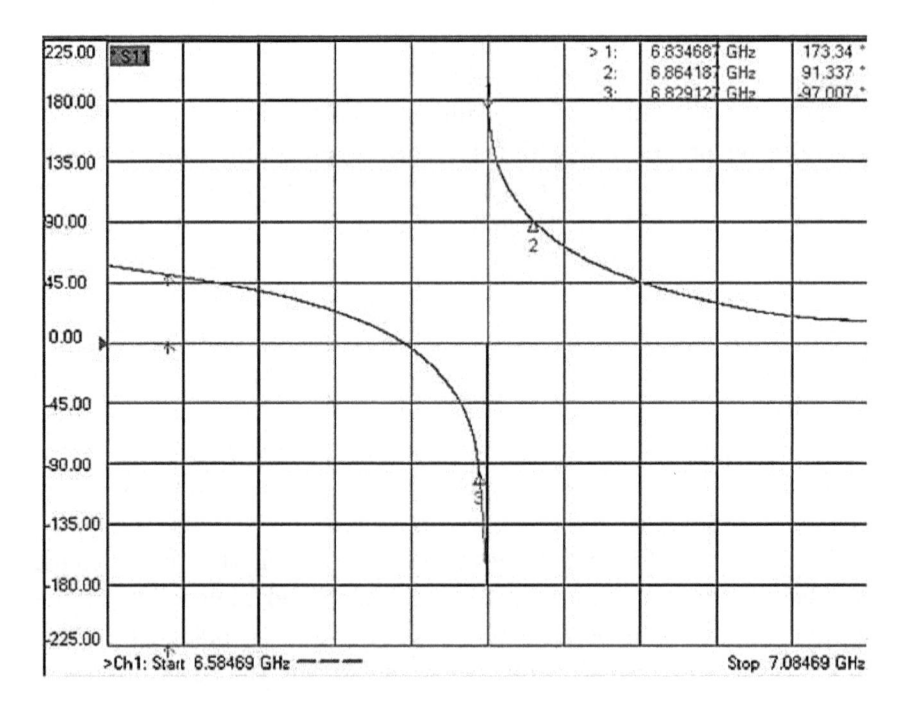

Fig. 6.6 Measured data of $|S_{11}|$ using a 6.834 GHz cavity.

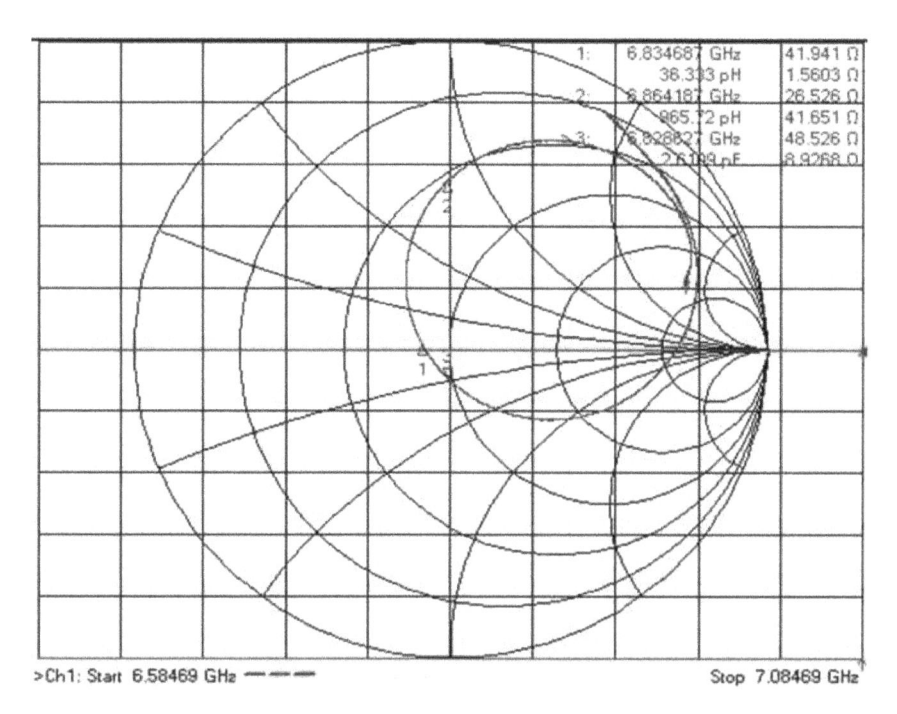

Fig. 6.7 Measured Q circle data of the input impedance of the 6.834 GHz cavity, plotted on the Smith Chart to demonstrate the tuning effect of the variable tunner. The resonant frequency f_L is indicated by 1.

the centre point (normalized to Z_0) at f_0. The plot on the magnitude in Fig. 6.6 reveals a deep null at this matched condition, approaching the asymptotic value of $\Gamma_{in} = 0$ (or $-\infty$ in dB). In Fig. 6.8, we see the dramatic improvement in the phase response of the cavity when the matched condition is achieved. This is also indicative of the discriminator curve obtainable with this matching technique. The measured characteristics of the cavity in Fig. 6.6, show the S_{11} response of the resonator for a frequency sweep from 6.5 MHz to 7 GHz.

This result establishes that the resonant frequency of the dominant mode of the cavity is 6.834 GHz and input return loss is greater than -20 dB. This is 3 MHz below the resonant frequency predicted by HFFS simulation. This is in good agreement. However, the minor difference may be for the reasons: (i) somewhat different practical values of the dielectric constant than 2.1 for Teflon,

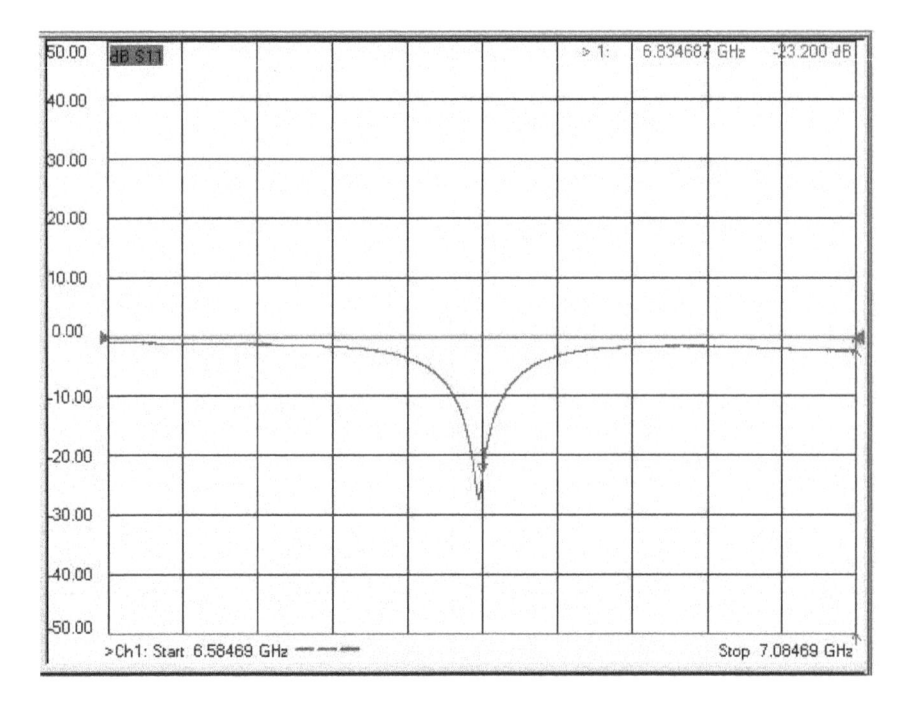

Fig. 6.8 Measured phase of S_{11} for a 6.834 GHz cavity.

3.8 for the absorption glass cell and 4.4 for FR4 PCB considered in the HFFS simulated model. (ii) The dielectric material is not quite homogeneous and perfect in nature. (iii) Input/output ports in the HFFS simulated model may not be same as in the fabricated cavity.

The frequency response of TE_{111} mode can also be used for Q determination of the resonator. The HFFS simulated and experimental frequency responses for determining Q are shown in Figs. 6.9 and 6.10, respectively. The actual resonant frequency and Q of the cavity are determined experimentally using a network analyser. The measured characteristics of the cavity in Fig. 6.10 show the S_{21} response of the resonator for a frequency sweep from 6.6 GHz to 7.0 GHz. It is evident from these figures, that this resonator has relatively low Q value of about 225. From HFFS simulation and the measured Q of the cavity, including glass cell, Teflon ring and photodiode PCB with mounted photodiode, is about 200. The Rb vapour has loss tangent

Design Simulation and Development of Microwave Cavity 181

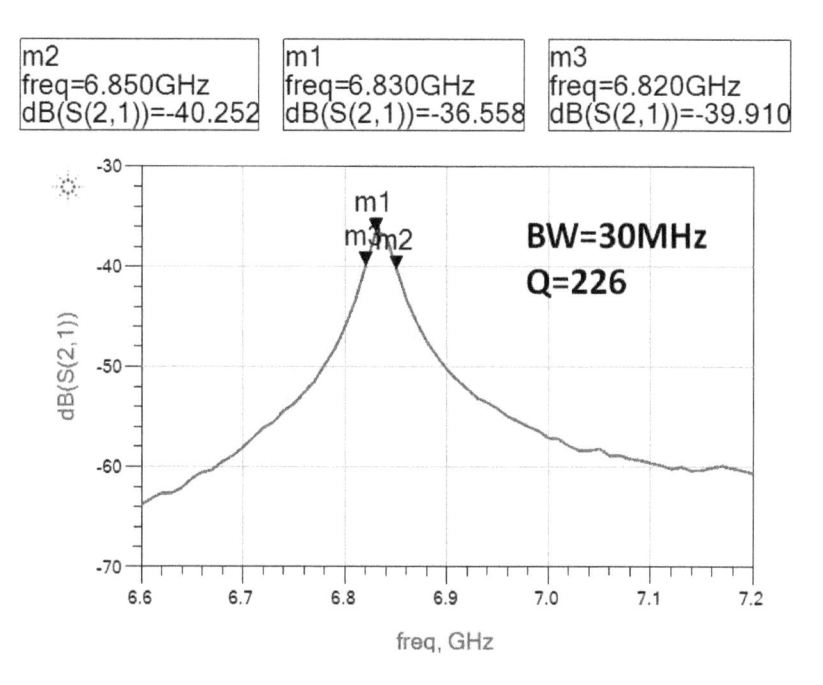

m2	m1	m3
freq=6.850GHz	freq=6.830GHz	freq=6.820GHz
dB(S(2,1))=-40.252	dB(S(2,1))=-36.558	dB(S(2,1))=-39.910

Fig. 6.9 S_{21} transmission as simulated by HFFS with Q=226 at f=6.834 GHz.

tan $\delta = 0.05$ and the simulations match with the theoretical results, the Eq. (6.5b). The achieved Q is slightly lower than the simulation results, because of the presence of a hole for the optical radiation to enter the cavity, the photodiode PCB cable. The photodiode itself adds to some losses due to their dielectric property. The reason for low Q is partially due to resonator being operated in TE_{111} mode. In this mode a large portion of the electric field propagates along the length of the resonator, causing high currents and the losses. The silver coatings are used to prevent the radiation losses. Finally, the loaded Q < 200 meets the passive atomic clock requirements [201] and ensures a negligible cavity pulling contribution to the atomic signal [202].

The resonator has small volume of the dielectric. The dielectric should be securely mounted between the absorption cell and the cavity wall. The advantage of this resonator is its compactness. In this cavity, very little magnetic field lies in the centre of the dielectric

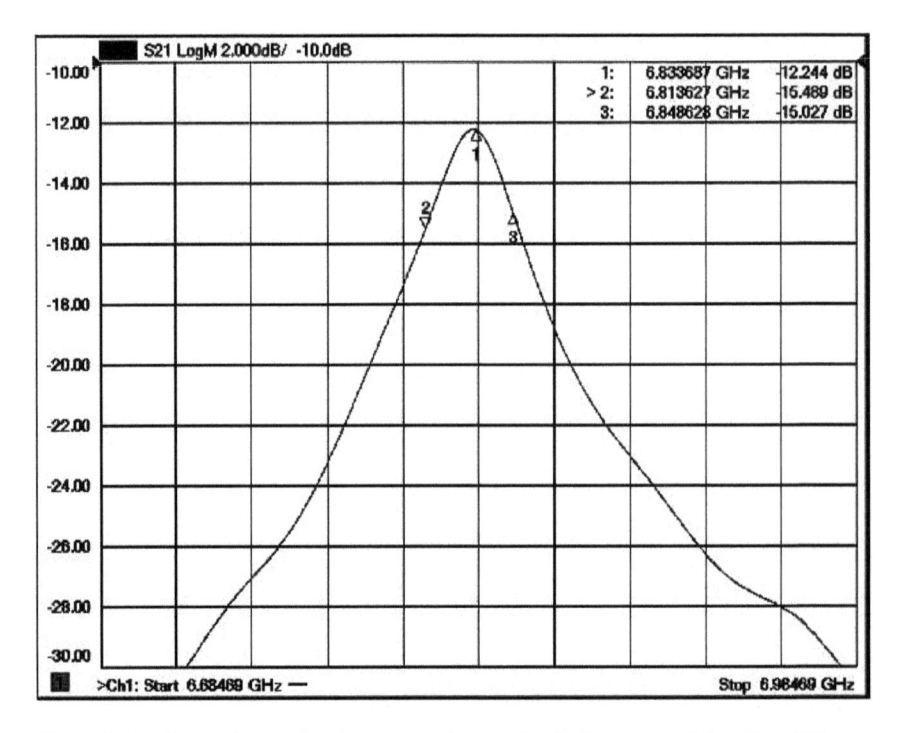

Fig. 6.10 Experimental determination of Q-factor, with Q = 194 at f = 6.834 GHz.

as shown in Fig. 6.4. It can be seen that the distribution of the microwave magnetic field and its direction is almost parallel to the direction of the DC magnetic field, as per the requirement of field free atomic ground state hyperfine transitions. We have developed a miniature microwave cavity with volume of about 23 cm^3. For the fabricated resonator with A or a = 15 mm, H or h = 33 mm, we are able to get good response and relatively high return loss at 6.834 GHz resonant frequency. The prototype microwave cavity assembled with photodiode PCB and the coupling loop are shown in Fig. 6.11.

6.5. Frequency shifts due to cavity pulling effect

The cavity pulling occurs if the eigen frequency of the microwave cavity is not tuned exactly to the atomic transition frequency. If

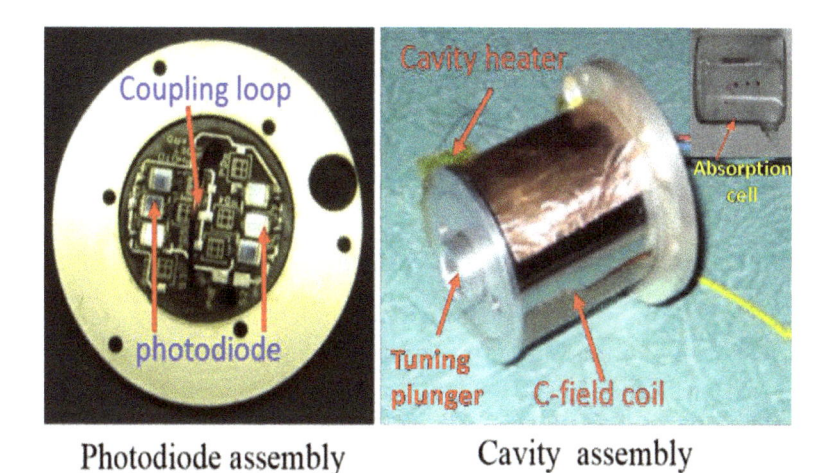

Fig. 6.11 Cavity assembly with photodiode detector and absorption cell.

the cavity is detuned from the resonance, the two frequencies are located on different positions of the resonance curve of the resonator, which leads to different excitation probabilities of the atoms. An approximation for the corresponding fractional frequency shift in Rb clocks is given as follows [201],

$$\frac{\Delta\nu}{\nu_0} \approx \left(\frac{Q_C}{Q_{at}}\right)\left(\frac{\alpha}{1+S}\right)\frac{\nu_C - \nu_0}{\nu_0}. \quad (6.8)$$

In Eq. (6.8), ν_c and ν_0 are the resonance frequencies of the cavity and atomic transition respectively. Q_c is the cavity loaded quality factor ($\approx$200), Q_{at} is the Rb line quality factor ($\approx 10^7$), α characterizes the operating conditions with respect to the oscillation threshold, which has the typical value of 10^{-2}, and S is the RF saturation factor $\approx$2 for the optimum discriminator slope. Thus, the cavity pulling effect can be reduced by employing a resonant cavity with low Q_c. The effect of the microwave power and temperature variations on the cavity pulling frequency shift can be minimized under reasonable RF power and temperature stabilization condition. If the clock is operated at the optimum power, (i.e., with $\pi/2$ excitation zone), the transition probability is largely independent of the power. The temperature of the cavity is regulated by adjusting the thermal control circuits, according to the calculated sensitivities of cavity frequency. However, it is clear that the cavity needs to be kept tuned to better than 200 kHz if a frequency stability better than 2×10^{-12} is required.

6.6. Frequency sensitivity of microwave cavity

From Eq. (6.6), we derive the coefficients of frequency sensitivity to the length, radius, and temperature of the cavity, as given in Eqs. (6.9)–(6.12):

$$\frac{df}{dl} = -\frac{1}{4}c\left(\frac{1}{4h^2} + \frac{1}{(3.41a)^2}\right)^{-\frac{1}{2}} \cdot l^{-3} = -93.6\,\text{MHz/mm}. \quad (6.9)$$

$$\frac{df}{dR} = -\frac{1}{3.41^2}c\left(\frac{1}{4h^2} + \frac{1}{(3.41a)^2}\right)^{-\frac{1}{2}} \cdot R^{-3} = -314\,\text{MHz/mm}. \quad (6.10)$$

$$l' = l(1 + \alpha t),\ R' = R(1 + \alpha t) \quad (6.11)$$

$$\frac{df}{dt} = -\frac{c^2\alpha}{f}\left(\frac{1}{4h^2} + \frac{1}{(3.41a)^2}\right)$$

$$= \{-176.5\,\text{kHz}/°\text{C}, \alpha_{\text{aluminum}} = 23.0 \times 10^{-6}/°\text{C}\}. \quad (6.12)$$

When the length or the radius increases by 1 mm, the cavity frequency decreases to 93.6 or 314 MHz, respectively. The frequency varies with the temperature at a rate of approximately 176.5 kHz/°C, for the aluminium cavity. The resonance frequency is set by the dimensions of the cavity and then it is regulated during the operation, by varying the temperature, through the thermal control circuit, according to the calculated sensitivities of cavity frequency. A linear shift of the atomic clock's transition frequency with the absorption cell temperature, contributes to the final stability limit of the clock. A cavity TC of 176 kHz/°C yields a RFS TC of about $2.43 \times 10^{-12}/°\text{C}$. The temperature coefficients of TE_{111} cavity are measured. In the experiments, we find that the shift in the resonance frequency of the TE_{111} cavity is very high and is proportional to

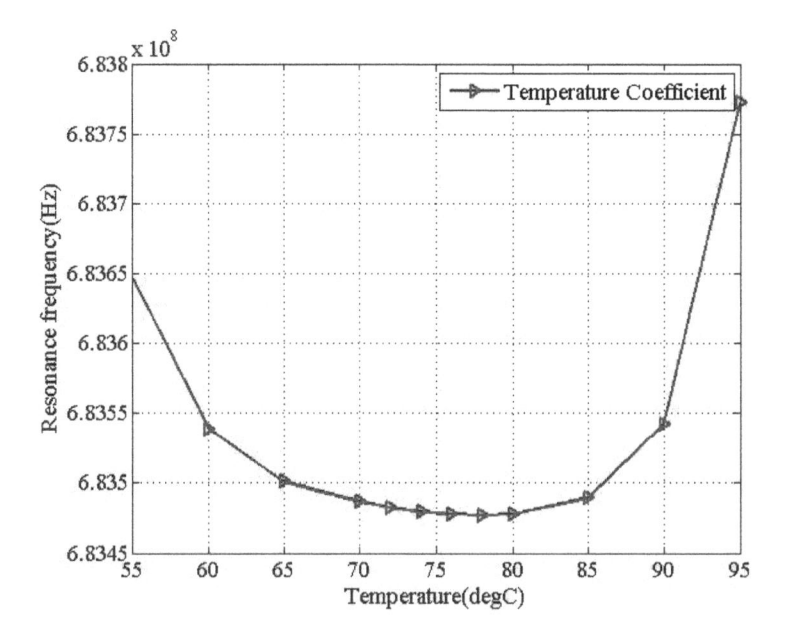

Fig. 6.12 Experimental measurement of TC for TE$_{111}$ cavity.

the temperature, which has a quite large range 55 to 70°C (TC = −51.5 kHz/°C) and 78 to 95°C (TC = 65.4 kHz/°C). But in the region of 70 to76°C ,which corresponds to operating temperatures of the cavity, the TC reduces to 3 kHz/°C. The appropriately chosen absorption cell temperature is around 72°C for which the clock TC limit is around 4×10^{-14}/°C and we have a realizable control of the cell temperature at the milli Kelvin level. The experimental results in range 55 to 95°C are shown in Fig. 6.12.

Summary

In this chapter, the HFFS simulation technique is applied to analyse the electromagnetic field in the cylindrical cavity, using reflection and transmission procedure. This process is helpful in the design of dielectric loaded cavity for the Rb atomic frequency standard. In the electromagnetic simulations, the fixed-dimension metallic circular cavity resonates in its dominant TE$_{111}$ mode at 6.834 GHz without dielectric load, which is slightly different than what is found by the calculation (i.e., 6.7 GHz). In the whole design process, resonant frequencies of the other higher order modes, quality factors and EM field distributions of dominant TE_{111} modes and TE_{011} are rigorously analysed, using the EM solver. The theoretical analysis agrees very well with simulated and experimental results. The versatile visualization of the field distributions gives much more insight into the behaviour of the mode fields. With the dielectric-loaded cylindrical cavity operated in the mode TE_{111}, a miniature cavity-cell assembly is successfully developed. As a result, the volume of the microwave cavity is only 23 cm^3. The resonant frequency and Q of the resonant cavity meet the requirements of the Rb atomic clocks.

The influences of the three dielectric materials i.e., absorption glass cell, Teflon ring, FR4 photodiode PCB along with the coupling loop dimension, on the resonance frequency and Q are analysed. We find that the inclusion of these dielectric materials into the cavity distorts the field lines considerably, degrades the Q and shifts the resonance frequency of the cavity. The level of EM field, an important parameter in the microwave cavity design, is investigated and experimentally verified. Experimental cavity measurements are compared with the analytical results and a close agreement is obtained with simulated values. An optimization method for cavity dimensions is also suggested. The obtained results with the loop feed, show that values of resonant frequencies mainly depend on wire dimensions, which are related to the loop impedances. The HFFS simulation technique gives information about the dimension and position of coupling loop, for achieving the best source matching in operating frequency range. In the present, case 8 mm loop length is chosen to excite the microwave cavity for the best results. The influence of the cavity pulling effect on the frequency stability of the passive Rb frequency standards is analysed. It is shown that the variation of microwave power and cavity temperature lead to an effect on the cavity pulling and the frequency shift, but can be minimized under reasonable RF power and temperature stabilization conditions. Therefore, a TE_{111} type of microwave cavity with low TC is developed. The temperature coefficient is measured to be 3 kHz/°C which results in a clock TC limit of around 4×10^{-14}/°C. It is achievable for a realizable control of the cell temperature at the milli Kelvin level. The analysis shows that the cavity pulling effect of the cavity can be neglected under reasonable temperature stabilization condition. The improvement in cavity TC is possible due to the compensation of the positive TC of the dielectric ring in the cavity to the negative TC of the metal part of the cavity.

Chapter 7

Design and Simulation of Analog Lock-in Amplifier

> The Rb Physics package generates the atomic resonance or clock signal. The quality of this error signal decides the overall performance of the Rb atomic clock. This signal is processed by the lock-in amplifier or phase sensitive detector. The atomic resonance error signal is applied to electronic frequency control of VCXO to check its aging. In this chapter we discuss details of this critical component of Rb atomic clock.

7.1. Introduction

The lock-in amplifier or phase sensitive detector is used in detecting very weak signal embedded in noise. The atomic signal detection is a big challenge as these signals are extremely feeble. Similarly, in electronics and cryogenics, the lock-in amplifier or phase sensitive detector is used in component characterization, bridge networks, and measurements of the resistance of superconductors. This versatile device finds applications in a weak AC signal recovery instrument, vector voltmeter, phase meter, and noise measurement unit. The Lock-in amplifiers (LIAs) are well known and widely used systems in physical and chemical sensing and materials spectroscopy applications [203–210] and widely used for frequency stabilization techniques in atomic clock application. It is an integral part of all atomic clocks in processing and utilizing weak atomic signal.

In the case of passive Rb atomic clocks, when the light passes through the Rb vapour absorption cell and resonant RF probes the atomic resonance, an error clock signal is obtained. Which contains information about the g.s hyperfine transition and is detected by a photodiode. The atomic resonance line shape can be represented by a Lorentzian, when the RF frequency, with low frequency or phase modulation, is exactly on the resonance, a null error signal is obtained. Any change in RF frequency from the resonance, causes the absorption signal to decrease. A small phase or frequency modulation of the RF frequency is done. After phase sensitive detection, a dc error signal is generated, which is proportional to the first derivative of the absorption signal. This dc error signal, when applied to EFC of VCXO, continuously forces the RF frequency back onto the peak of a hyperfine atomic transition. As the RF frequency naturally drifts in one direction or the other, away from the peak, the error signal develops and it is used to correct changes in the RF frequency. Since it has a sloping section crossing zero at the resonant frequency.

For locking the frequency of our RF source to the g.s hyperfine atomic transition frequency of ^{87}Rb, the only tuning parameter is the error control voltage for the VCXO/OCXO. The error signal fed to the VCXO/OCXO, continuously corrects the output frequency of VCXO/OCXO, so the RF frequency stays locked onto the point of resonance. At the resonance there is no first harmonic signal, as measured by the lock-in amplifier. As a result, the detected signal by a photodiode has only a frequency component which is twice the modulation frequency. This keeps the RF source stable at a hyperfine atomic transition of ^{87}Rb. Therefore, the purpose of the lock-in amplifier is to provide a DC output voltage, known as the error signal, which is proportional to the amount of first harmonic signal present in the lock-in input signal.

The slowly varying detected optical signals are extremely weak in amplitude, even smaller than the coupled electrical noise level, common in many optical measurement situations. It is often difficult to detect such signal due to the presence of high-level, low frequency optical interference and low frequency (1/f) noise. In these situations, the typical frequency-selective based filtering techniques cannot be employed to extract the desired signal from the noisy background environment. Because of the fact that this technique is still vulnerable to the noise that exists in surrounding light and also in power network. One should look for the special techniques. A solution to this problem is to use a lock-in amplifier [211–212] based detection, which uses the phase-sensitive detection technique (PSD) [213–214] to single out the desired component out of the signal at a specific reference frequency and phase. The noisy signals at frequencies other than the reference frequency, are rejected and do not affect the measurement. In short, LIAs improve the signal-to-noise ratio (SNR) of the signal and are the preferred choice in applications where the desired output signal is usually embedded within considerable noise. In many situations, the noise power is many times more than that of the desired signal, which are generally very weak [215–217].

In this chapter, we describe how the detection is accomplished with the lock-in amplifier. We design a simple, re-configurable, high performance, analog LIA, that is small and simple enough to be included as part of a larger instrument. In the Rb atomic clock, PSD is operated at 137 Hz and is capable of extracting signals as small as 100 nV from the ambient noise. The developed LIA, includes a high gain photodiode array to convert the incident optical signals into electrical current signals. The photo-current is then amplified and converted into a voltage signal by a low-noise Trans-impedance amplifier (TIA). A low-noise amplifier (LNA) and a band-pass filter first process the signals. Then, the signal is multiplied, with a known reference signal of the same frequency as the input signal, by a mixer. The multiplied product signal is then converted down to a low-frequency signal from which, the amplitude and phase of the desired signal are extracted. A low-pass filter filters out the dc component. The proposed LIA consumes an average power of 1.3 W with a ±15 V power supply. The LIA enables the recovery of signals 60 dB below the noise level (a voltage ratio of 10^3). In accepted terminology, this corresponds to a dynamic range of 140 dB. In this chapter; we provide a simple physical explanation and describe the design and measurement of an integrated lock-in amplifier, based on a suitable mathematical model. This work is outlined as follows. In Sec. 7.2, we explains how the "lock in" produces the desired feedback signal and describe the shape of the feedback discriminator signal, using a simple open loop experiment. Section 7.3 briefly describes the analog lock-in architecture and its main building blocks. In Sec. 7.4 analyses the design and implementation of the major sub-circuits of LIA. In Sec. 7.5, the SNR of integrated Physics package measurements, using lock-in instrument, RF synthesizer and spectrum analyser are presented. The summary in Sec. 7.6 gives the analysis on the lock-in amplifier.

7.2. Physical description and experimental characterization of discriminator signal

Phase-sensitive detection provides a sensitive feedback signal for the local oscillator (i.e., OCXO), that is typically locked to the Rb g.s hyperfine transition, to make a functioning clock. Typically, a low frequency sinusoidal signal is superimposed on the microwave source, leading to sine wave modulated, or dither, the RF frequency Ω, producing a voltage signal V (t) = V ($\omega(t)$) $\approx$ V [$\omega_{center} + \Delta\omega \cos(\Omega t)$]. If $\beta = \Delta\omega/\Omega \rangle\rangle$ 1, then the voltage signal behaves as if the RF frequency were slowly oscillating back and forth.

A lock-in amplifier with reference frequency Ω can then produce an error signal $\varepsilon(\omega)$ that is the Fourier component of V (t) at frequency Ω. Expanding V (ω) $\approx V_0 - A(\omega - \omega_0)^2$ for ω near the peak frequency ω_0, we have

$$V(t) = V[\omega_{center} + \Delta\omega \cos(\Omega t)]$$

$$\approx V(\omega_{center}) + \frac{dV}{dw}(\omega_{center}) \cdot \Delta\omega \cos(\Omega t) + \cdots$$

and so the Fourier component is

$$\varepsilon(\omega) \sim \frac{dV}{dw}(\omega) \Delta w$$

$$\sim 2A \Delta\omega (\omega - \omega_0).$$

This error signal has the desired properties that $\varepsilon(\omega_0) = 0$ and $\{d\varepsilon/d\omega\} \neq 0$ for $\omega = \omega_0$; thus it can be used in a feedback loop to lock the RF frequency at ω_0.

When the lock-in amplifier deals with the V (t) signal and the reference signal, we get the differential signal. After demodulation, the almost symmetric Rb resonance is transformed into an approximately anti-symmetric discriminator signal. Near an exactly symmetric Rb resonance, this discriminator signal is proportional to the frequency difference between the local oscillator and the centre of the Rb resonance. And thus, one can lock the local oscillator exactly to the Rb resonance, by measuring the microwave frequency corresponding to the zero crossing of the discriminator signal. However, any asymmetry in the Rb resonance line shape, shifts the zero crossing of the discriminator signal, with a magnitude dependent on the slow-phase-modulation parameters. The instability in these modulation parameters thus induce instability in the Rb clock's microwave frequency. Figure 7.1 shows a schematic of our experimental setup. We derive the dispersion like error signal needed for the Rb clock, by phase modulating the output from an SMP04 R&S signal generator tuned in the vicinity of the 6.834 GHz ground state hyperfine transition frequency ω_0 of Rb atoms. A photodiode detects the absorption of the light by the Rb absorption cell. The absorption signal is fed

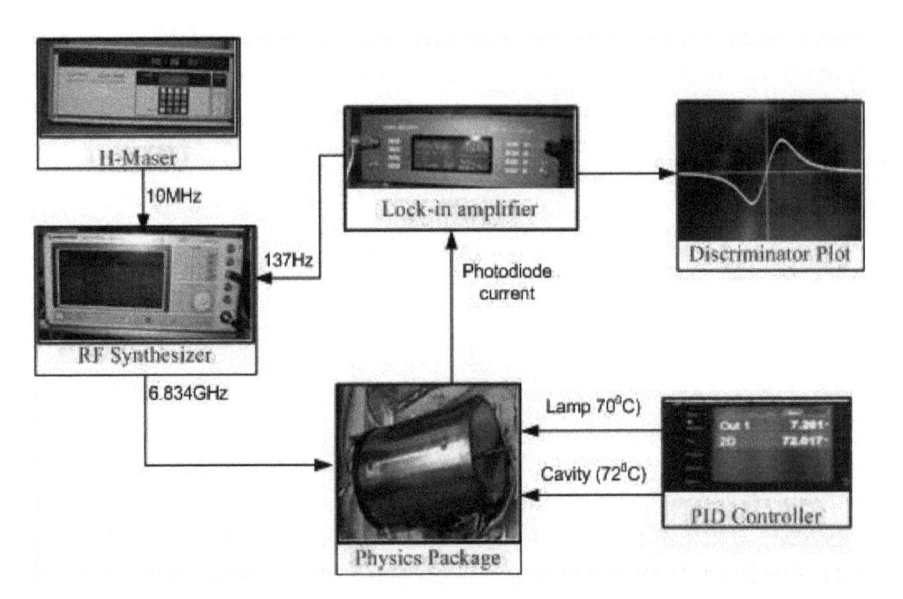

Fig. 7.1 Experimental setup for Physics package discriminator signal characterization.

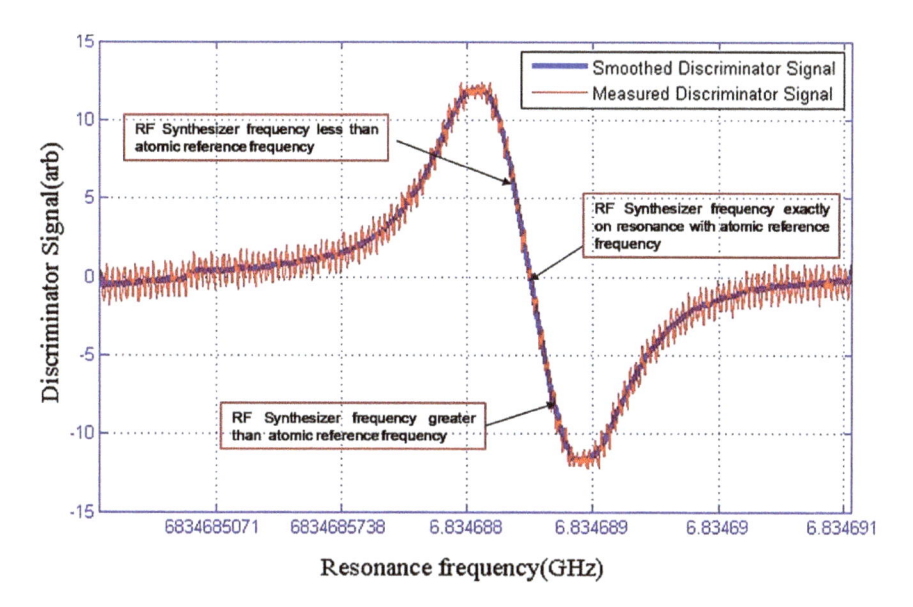

Fig. 7.2 Discriminator signal at the output of lock-in amplifier shows the error signal that is derived from the absorption signal by the lock-in amplifier at different RF frequencies.

into the lock-in amplifier instrument, which provides an output signal, that is directly proportional to the amount of ω_m (modulation frequency) signal present in the absorption signal, while ignoring all other frequency components. We use phase-sensitive detection to convert the approximately symmetric Rb resonance of the double resonance clock transition into a dispersive like resonance as shown in Fig. 7.2.

We apply a slow phase modulation to the microwave signal, driving the microwave cavity and then demodulate the corresponding slow variations in the photodetector current, using a lock-in amplifier. In the Rb clock, a feedback loop locks the external oscillator, VCXO/OCXO to the zero crossing of the anti-symmetric modulated signal. Here, rather than closing the loop in the feedback system, we measure the frequency of the zero crossing. By fitting a line through the central part of the dispersive like resonance, relative to a frequency source phase locked to a H-maser. As a result, the output

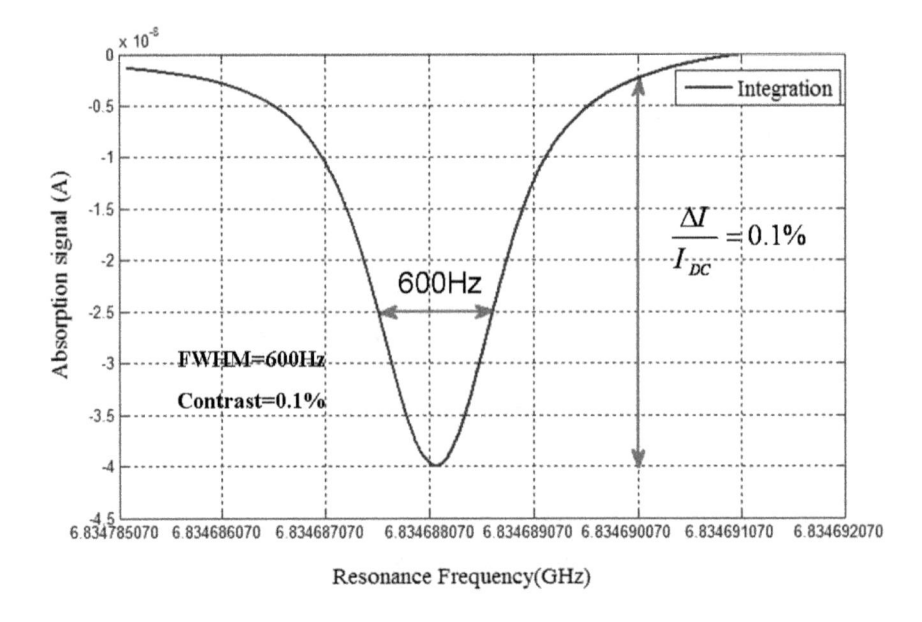

Fig. 7.3 Absorption-resonance signals.

signal of the lock-in amplifier serves as an error signal, that is proportional to the first derivative of the absorption signal. It corrects for small changes in the RF frequency. A typical absorption signal of the clock transition from the clock setup is shown in Fig. 7.3. Experimental parameters are: an absorption cell at the temperature 72°C, a RF power −19 dBm to the Physics package and a static magnetic C-field of 244 mG, applied in the direction of the optical path, to split the Zeeman sub-levels and detect only the $m_f = 0 \leftrightarrow 0$, clock transition.

The absorption signal has a contrast of 0.1% and a narrow line width of 600 Hz, resulting in an error signal with an excellent discriminator slope of 271pA/Hz. If the RF frequency drifts above the desired value, we have the negative feedback signal. This changes the control voltage so as to correct the RF frequency. Similarly, if the RF frequency drifts below the desired value, we get the positive feedback signal. In our case, the desired frequency corresponds to the peak of an atomic resonance. Therefore, when the RF frequency

ω_{RF} exactly matches the external reference frequency ω_o, no error signal is generated [211]. However, a drift to higher or lower of the resonance frequency, results in an error signal that is directly proportional to the amount of frequency drift. The error signal therefore provides a continuous restoring signal that can be used to keep the RF frequency locked to the atomic reference frequency.

7.3. Functional aspects of the lock-in amplifier

The local oscillator correction servo is implemented with a compact; low-power analog lock-in amplifier system, which is composed of five major parts. It consists of a Trans-impedance amplifier, a high gain amplifier, two BPF, a mixer, a differential amplifier and a low pass filter. Figure 7.4 shows the lock-in amplifier schematic block diagram. The circuit operates at a dual $+/-15$ V supply [218–219], which is discussed in this chapter.

The detected photo-current from the Physics package photodiode is processed through trans-impedance amplifier and a high gain

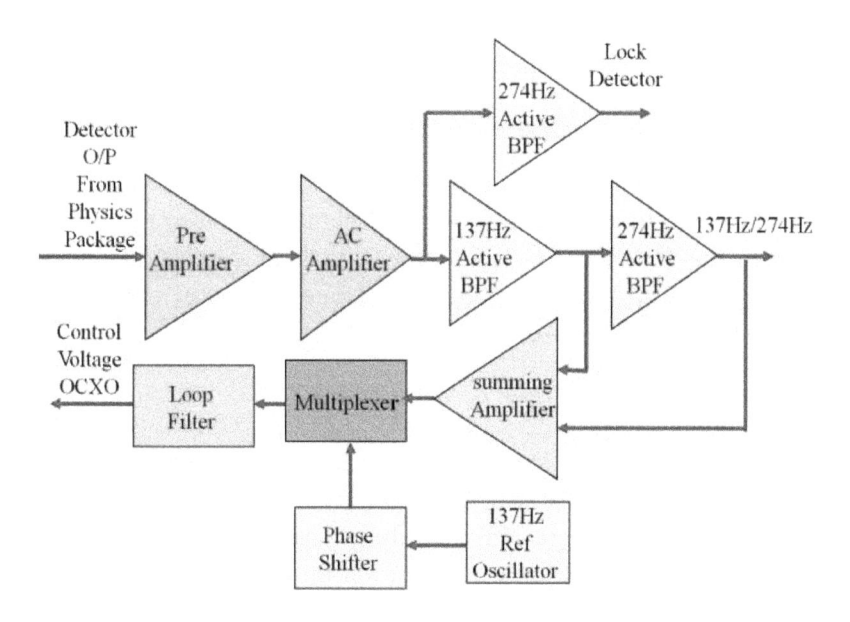

Fig. 7.4 Block diagram of the analog lock-in amplifier.

AC amplifier, its output voltage, which contains the amplitude and phase information of the signal, is filtered by a band-pass filter, to remove the harmonic interferences. The resulting AC signal is sent to the input port of a phase-sensitive detector (PSD), which has in the present case, the 137 Hz modulation signal, with the variable phase shift, as its reference. The modulation signal for the RF source and also for the reference of the PSD, is generated by a single chip in a self-oscillation configuration. Each signal is sent to a flip-flop, that divides the frequency by a factor of two. The output from the flip-flop in the LO channel and for the lock-in reference is sent through a band-pass active filter, which eliminates the DC component. The remaining AC signal is sent to the lock-in reference port and LO input port through an active phase shifter. The output of the PSD is filtered with a low-pass RC filter to eliminate the modulation component and then integrated to provide the correction signal. This correction signal is sent to the EFC of the VCXO/OCXO, to correct its frequency. All components of this locking system are surface-mount devices on printed circuit boards. The three boards have a volume of 6.3 cm^3 and all components together dissipate a total of about 1.2 W power.

7.4. Design and implementation of the lock-in amplifier

The schematic block diagram of the designed optoelectronic LIA is shown in Fig. 7.4. The on-chip photodiode array detects the incident optical signals and generates a proportional electrical current. The photocurrent signal can be expressed as:

$$I(t) = I_0 \cos(w_0 t) + I_D, \qquad (7.1)$$

here I_0 is the peak current intensity, I_D is the phototransistor dark current, and w_0 is the sinusoidal modulation angular frequency. The current signal is converted into a voltage signal through a high-gain trans-impedance amplifier where the dc and ac components of the current are amplified. The high-pass filter after the TIA removes the undesired dc offset voltage. The output signal is then processed by a band pass filter to remove the noise and the interference with tones other than w_0. The high-pass filter after the band pass filter then removes the dc offset voltage introduced by the band pass filter. The signal, after the second high-pass filter and the reference signal are sent to a mixer circuit. The mixer output signal can be expressed as:

$$V(t) = I_0 A + I_0 A \cos 2w_0 t, \qquad (7.2)$$

here A is the combined gain of the TIA, high-pass filter, band-pass filter, and of the mixer. The signal $V(t)$ consists of two tones located at the dc and $2w_0$, respectively. A low-pass filter is added after the mixer to remove the tone at $2w_0$. Therefore, the low-pass filter output signal V_{out}, a DC voltage, is directly proportional to the photocurrent I_0, where I_0 is proportional to the incident optical power. The sub-sections describe the operation of each functional block in the LIA.

7.4.1. Dither (137 Hz) and input (absorption signal from Physics package) signal

The lock-in amplifier requires a dither signal and an input signal. The dither signal defines the frequency, 137 Hz, at which the lock-in amplifier recovers signals from the input channel. In general, the dither signal stimulates an experiment, which responds at the same frequency. This response is the input signal for the lock-in amplifier. In our experiment, the dither signal is provided by an external sine-wave generator. The input signal comes from the atomic absorption spectrum of the Rb atom. Both signals have the same frequency and phase. The dither signal goes through a timer. The timer generates the reference square wave and it is converted to sine wave by using an active low-pass filter. This sine wave signal is used as a reference for driving the mixer circuit.

7.4.2. Photodiode detector and trans-impedance amplifier (TIA)

The incident light is detected by an array of photodiodes. The area of each photodiode is 1 mm × 1 mm, and the overall area of the photodiode array is 10 mm × 10 mm. The photodiode array is so designed that all the anode terminals are connected to the input of the TIA and the cathode terminals are connected to the ground.

The noise in the clock is dominated by the TIA, which is the first component connected to the output of photodiode detector. The TIA is based on a unity gain stable voltage feedback amplifier, with the bandwidth of the order of 8 MHz. The devices from several manufacturers meet this requirement. We use the Analog Devices OP-27, as these op-amps give the best stability and low noise. A schematic diagram of the TIA is shown in Fig. 7.5. The device capacitance C_1, at the input port of the circuit results in a second-order response, given by:

$$\frac{V_0}{I} = \frac{R_2}{1 + 2\xi(s/\omega_n) + (s/\omega_n)^2}, \qquad (7.3)$$

which relates the output voltage Vo to the diode current I. The damping factor ξ and the un-damped natural frequency ω_n are given by $\omega_n = [Ao/\{(C_1+C_2)R_2\tau_o\}]^{1/2}$ and $\xi = 0.5\ \omega_n\ \{(C_1+C_2)R_2/Ao + C_2R_2 + \tau_o/A_o\}$. Equation (7.3) also gives the ratio of the voltage

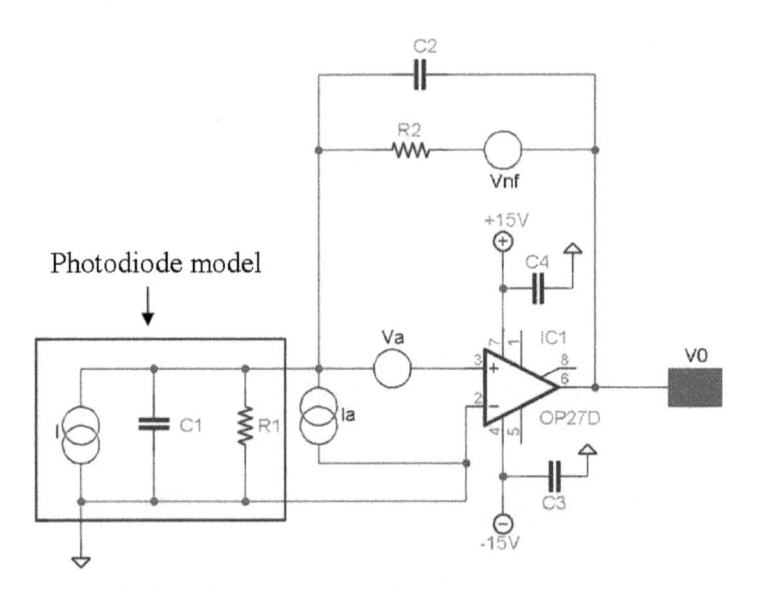

Fig. 7.5 The circuit details of TIA.

noise output V_{no} of the TIA to the photodiode noise current I_n. In arriving at this result, it is assumed that $R_1/R_2 \gg 1$. The op-amp open loop gain can be approximated by a first-order system behaviour, $A_v = A_o/(1 + s\tau_o)$. The open-loop dc gain A_o is much greater than unity. The capacitor C_2 is introduced to control the circuit damping. In the absence of C_2, the damping term in Eq. (7.3) is small and the gain peaks near the circuit's un-damped natural frequency. The resistance R_2 governs the trans-impedance gain. C_1, C2, R_2 and the amplifier gain-bandwidth determine the useful bandwidth of the system. The resistance R_2 and capacitance C_2 are chosen as 100 kΩ and 100 pF respectively, to give a reasonable compromise for maximizing both the trans-impedance gain and bandwidth. In reality, the model based on the first-order op-amp behaviour underestimates problems of the stability and gain peaking, for two reasons. Firstly, the op-amp is a higher order system and introduces phase shifts, howsoever small, in addition to that is expected of a first-order system. Secondly, the layout of the circuit and the physical nature of its components introduce parasitic elements, particularly, inductance, which is not considered in the model. The simulated responses of the TIA are shown for five different values of C_2 in Fig. 7.6(a).

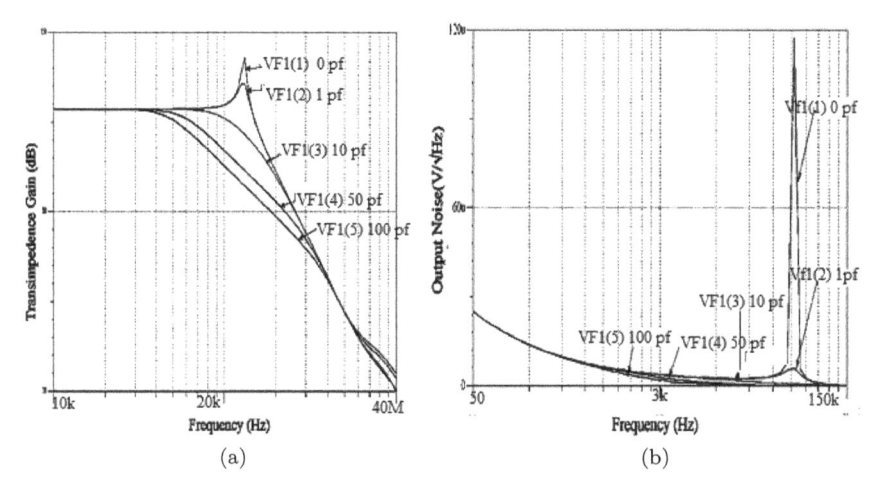

Fig. 7.6 (a) TIA frequency response for $Cp = 0 - 100\,\text{pF}$ (b) simulation of the TIA output noise spectrum with $Cp = 0 - 100\,\text{pF}$.

7.4.3. High-pass filter (HPF)

The photodiode output currents include the DC and AC components, both the components are amplified by the TIA. While the AC component does not affect the output DC voltage, the DC component leads to an undesired DC offset voltage at the output, which imposes an unintended biasing voltage on the signal-processing stages. Therefore, a high-pass filter is added after the TIA to remove the undesired DC offset voltage. The high-pass filter shown in Fig. 7.7 blocks any DC offset voltage, that the input signal may have. The filter has a very low cut-off frequency, $f_{3dB} = 1.6$ Hz to avoid reducing the amplitude of the AC signal. The reference frequency used in the experiment is 137 Hz, which is well above the cut-off frequency.

7.4.4. Band-pass filter (BPF)

The band-pass filters should have a large tuning range, and the Q must be high enough to filter out unwanted signals and noise. A special technique is used for good tuning ability and higher gain. This topology of the band-pass filter is called multiple-feedback filter or Deliyannis-Friend filter [220–221] as shown in Fig. 7.8. It exploits the full open loop gain and also refers to as infinite-gain filters.

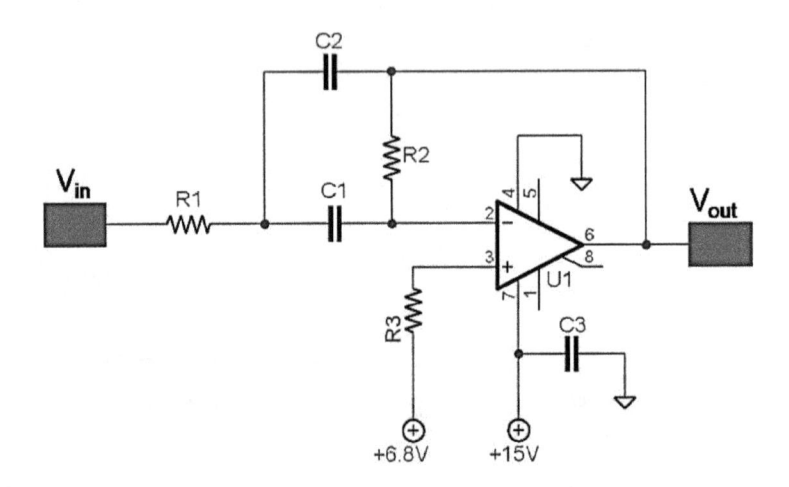

Fig. 7.7 Deliyannis-Friend filter schematic.

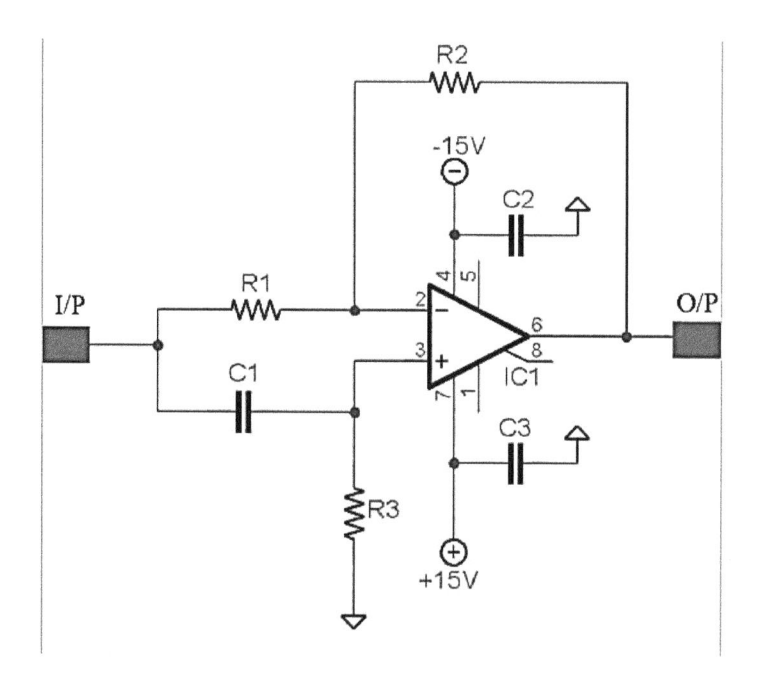

Design and Simulation of Analog Lock-in Amplifier 201

Fig. 7.8 Schematic of phase shifter circuit.

A Deliyannis-Friend filter band pass filter is simulated to check the frequency response. The circuit transfer function H(s) shows that the filter has a second order band-pass response with two poles and a zero at the centre frequency:

$$H(jw) = \frac{-jwR_2C_2}{1 - w^2R_1R_2C_1C_2 + jwR_1(C_1 + C_2)}. \quad (7.4)$$

The center frequency and Q are given by:

$$w_0 = \frac{1}{\sqrt{R_1R_2C_1C_2}}, \quad (7.5)$$

$$Q = \frac{\sqrt{\frac{R_1}{R_2}}}{\sqrt{\frac{C_1}{C_2}} + \sqrt{\frac{C_2}{C_1}}}. \quad (7.6)$$

The design challenge of the band-pass filter is to pinpoint the centre frequency within the required operating frequency range, without using unreasonably large capacitors while maintaining a good quality factor Q. According to Eqs. (7.5) and (7.6), the central frequency w_0 can be tuned by varying R_1 and R_2, keeping $C_1 = C_2 = C$. The quality factor Q can be improved by increasing the ratio of R_2/R_1.

7.4.5. *Phase shifter*

The dither signal typically, has a phase shift with respect to the input signal, arising from the response of the RF system and the optical system (Physics package). The control over the relative phase difference between the dither and the input signals is crucial in the generation of the final output signal of the lock-in amplifier. We construct a phase shifter, that makes it possible to change the phase of the dither signal, Fig. 7.8. In experiments, however, manual phase-shift modulation is a must as the phase of reference signal cannot track the phase of input signal automatically. To know whether these two phases are consistent, it is necessary to observe the signal manually, which may inevitably cause errors. Moreover, as different errors may be produced every time, when making observations, the measurement results are less comparable and the whole system is less precise and reliable. When considering above problems, the software TINA (Toolkit for Interactive Network Analysis) is used for modelling and simulating the phase shifter circuit, so as to optimize the design of traditional lock-in amplifiers.

As displayed in Fig. 7.8 the phase shifting circuit consists of a 300 kΩ potentiometer (R3), which allows the change in the phase of the dither signal. Figure 7.9 demonstrates the simulated results on the dependence of phase shift on resistance, R3.

The phase shifting circuit allows the phase variation of the dither signal. Although our phase shifters have range from 10° to 160°, rather than the full 0° to 360°, but this range is sufficient.

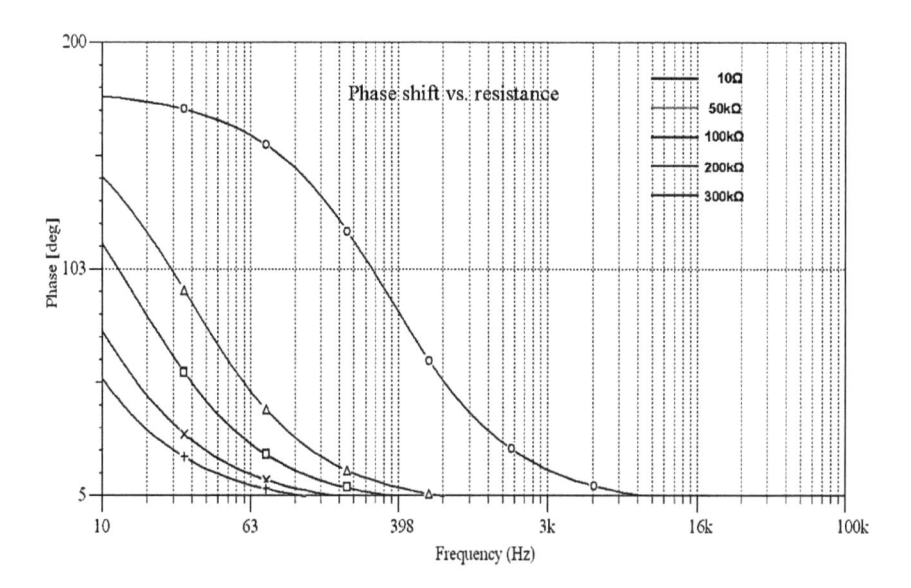

Fig. 7.9 Simulated results of Phase shift vs resistance.

7.4.6. Mixer (phase sensitive detector)

In the centre of our lock-in amplifier is a frequency mixer which multiplies the amplified input signal with the phase-shifted reference signal. The mixer or phase sensitive detector is a linear multiplier, or analog switch, whose output is the product of two sinusoidal inputs V_{in} and V_{ref}. When two waveforms, V_{in} and V_{ref} shown in Fig. 7.10, are multiplied together, the resulting $V_{out}(t)$ contains the sum and difference frequencies, $(\omega_{in} + \omega_{ref})$ and $(\omega_{in} - \omega_{ref})$. In our case, since the input signal and the dither signal are at the same frequency ($\omega_{in} = \omega_{ref}$), there are two components in the synchronous mixer output: a double frequency component and a phase-sensitive DC term. The DC voltage is at its maximum value when the input and dither signals are in phase and equals, $V_1 = V_2$. Figure 7.11 show the outputs of the mixer, when RF synthesizer frequency is below or above or on the resonance.

The design of the mixer, the fundamental circuit of the lock-in amplifier, is based on considering its operation in single supply mode. The proposed structure, shown in Fig. 7.10, consists three

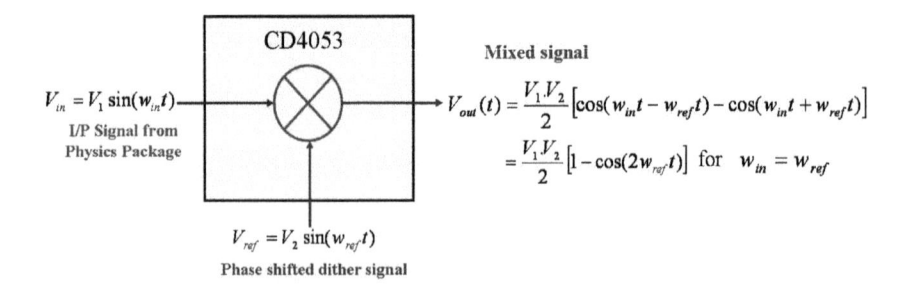

Fig. 7.10 Basic block diagram of mixer circuit.

Mixer diagram:
- $V_{in} = V_1 \sin(w_{in}t)$ — I/P Signal from Physics Package
- $V_{ref} = V_2 \sin(w_{ref}t)$ — Phase shifted dither signal
- CD4053 mixer output:

$$V_{out}(t) = \frac{V_1 V_2}{2}\left[\cos(w_{in}t - w_{ref}t) - \cos(w_{in}t + w_{ref}t)\right]$$

$$= \frac{V_1 V_2}{2}\left[1 - \cos(2w_{ref}t)\right] \text{ for } w_{in} = w_{ref}$$

CD4053 SPDT switches, so that it acts like a mixer with unity gain. The CD4053BC is a triple 2-channel multiplexer having three separate digital control inputs and an inhibit input. Each control input selects one of a pair of channels, which are connected in a single-pole double-throw configuration. The CD4053 switch has a superior isolation characteristic. It has low "ON" resistance: 80 Ω (typ.) over entire signal input range for $V_{DD} - V_{EE} = 15$ V and high "OFF" resistance with current leakage of ± 10 pA (typ.) at $V_{DD} - V_{EE} = 10$ V. A precise sine or square wave is required to drive the switch. The schematic of the mixer is shown in Fig. 7.10. The mixer outputs for RF being above or below or at resonance are shown in the Fig. 7.11.

7.4.7. Low-pass filter (LPF)

The clock signal is DC signal. It is necessary that the sum frequency is eliminated and only DC is allowed. As indicated in Fig. 7.8, in addition to the DC component another component at twice the frequency of the dither signal ($2\omega_{ref}$) results from the mixing. This component is required to be eliminated so that the output of the lock-in amplifier is only the DC voltage. A low-pass filter with a corner frequency lower than f_o is used to eliminate the $2\omega_{ref}$ component in the mixer output. It has a cut-off frequency of 1.6 Hz, which may be varied, so as to avoid attenuating the DC voltage. It reduces the amplitude of signals with frequencies higher than the cut-off frequency, f_{3dB}, at which 70% of the signal voltage (half the initial power) is allowed

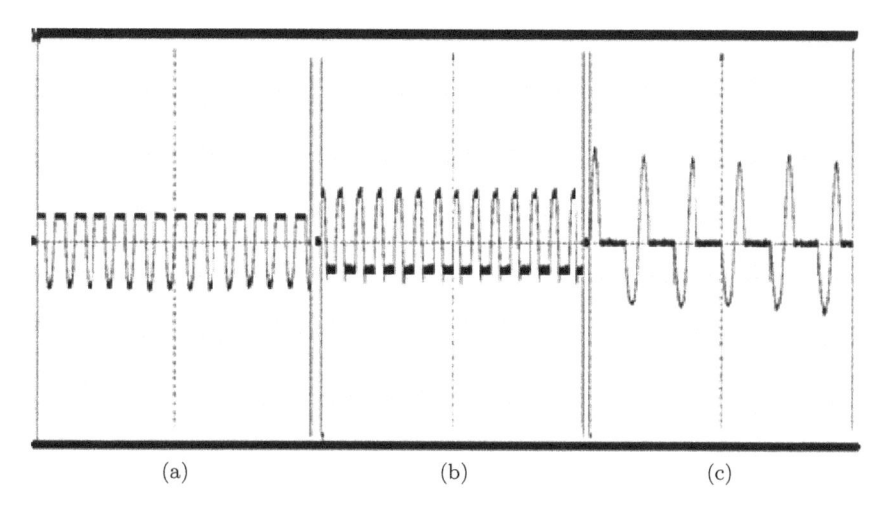

Fig. 7.11 Mixer output when RF (a) above (b) below (c) equal to the resonance frequency.

through i.e.,

$$f_{3dB} = \frac{1}{2\pi RC} = \frac{1}{2\pi \tau}.$$

Here $\tau = RC$ is the time constant. In our circuit, we can vary the f_{3dB} of the low-pass filter by changing R (τ = 1ms to 3s and f_{3dB} = 50 mHz to 160 Hz). In addition to the $2\omega_{\text{ref}}$ signal, the mixer output also inevitably contains some noise components at many different frequencies. The majority of the noise components are also attenuated by the low-pass filter. By increasing the time constant, the noise components can be reduced further. However, in this process, the cut-off frequency is lowered and the DC voltage may be attenuated, which varies slowly with time. We, therefore, have a trade-off. The simulated bandwidth of the low-pass filter is 1.6 Hz, where R_1 and C_1 are 10 kΩ and 10 uF, respectively. At the output of the low-pass filter, we get the required DC error signal. Figure 7.12 shows the effect of the low-pass filter in eliminating the higher frequency component, $2\omega_{\text{ref}}$, of the mixer output in Fig. 7.12(b), leaving only the DC component, i.e., the clock error signal.

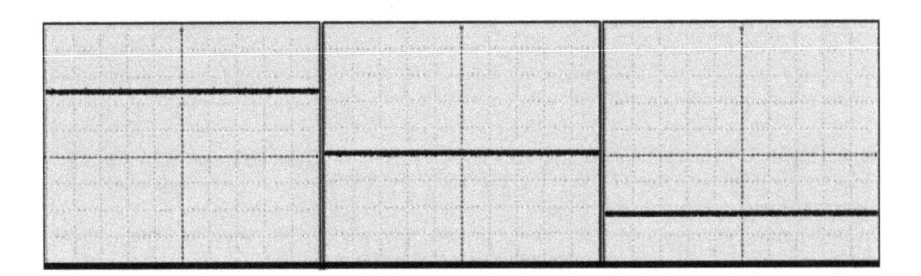

Fig. 7.12 The error signal after the LPF (a) error voltage is +4.9 V when $f_{RF} = f_{res} - 300$ Hz (b) error voltage is 0 V when $f_{RF} = f_{res}$ (c) error voltage is -4.4 V when $f_{RF} = f_{res} + 300$.

7.5. SNR measurement test setup

The measurement of SNR of the clock error signal from the Physics package is done, using Lock-in amplifier, RF synthesizer (10 MHz–40 GHz) and spectrum analyser. The designed analog lock-in amplifier's ability to recover information from the noisy signals is evaluated, using a 137 Hz sinusoidal signal. The SNR is defined as:

$$SNR = \frac{A_{SIGNAL}}{A_{NOISE}}.$$

The light is detected with a array photodiodes. Its intensity at the detector is a measure of the transparency of the Rb absorption cell, which is reduced, as explained earlier, on applying the microwave radiation at 6.835 GHz, the atomic resonance frequency. As the microwave radiation, with -18 dBm of microwave power is frequency modulation at 137 Hz, the signal detected at the photodiode and processed with a lock-in amplifier, produces a correction signal having a dispersion-like shape. This correction signal is the heart of the clock, producing a voltage that ties the output frequency of the OCXO/VCXO to the atomic transition. A block diagram of our SNR measurement system is shown in Fig. 7.13. It is shown the change in the lamp light transmitted through the Rb vapour, as the microwave signal is scanned across the 0–0 hyperfine resonance i.e., 6834.7 MHz. In the middle is the 2nd harmonic signal, that is the lock

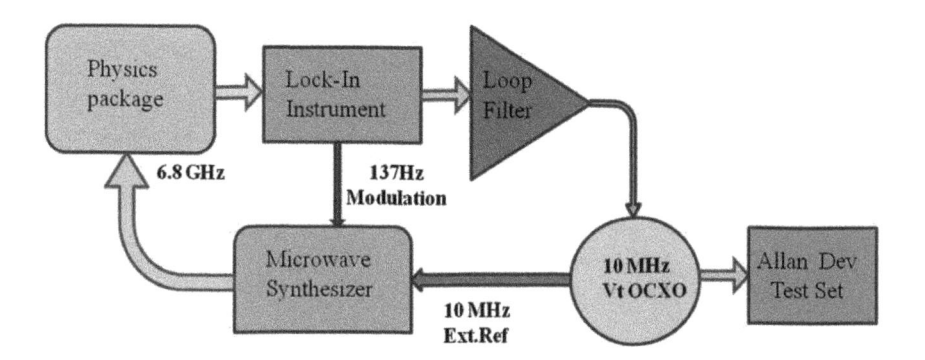

Fig. 7.13 Integrated test setup for SNR measurement.

point for the voltage-controlled-crystal oscillator (VCXO) to the Rb atom's ground state hyperfine transition. This signal is the maximum, when the microwave frequency is tuned exactly to the atomic resonance, and the strength of this is a measure of the 1st harmonic signal's slope on resonance. Since the slope of the 1st harmonic signal is a measure of the correction signal's ability to discriminate against small frequency variations of the microwave frequency and that of the VCXO. The 2nd harmonic signal is a measure of the error signal's quality. When microwave frequency is ± 300 Hz within the resonance width, away from atomic resonance, we get the maximum value of 1st harmonic signal. The 1st harmonic signal output is measured, with a spectrum analyser, as a function of the frequency. The results are shown in Fig. 7.14.

The results of Fig. 7.14 give the signal level at 137 Hz in a 1-Hz bandwidth. The frequency of 137 Hz is chosen in order to be well above the frequency, at which flicker noise [222–225] becomes significant. The measured results of servo section is shown in Table 7.1.

From Fig. 7.14 we get the SNR of clock signal from the Physics package and Fig. 7.15 gives the corresponding stability of the clock.

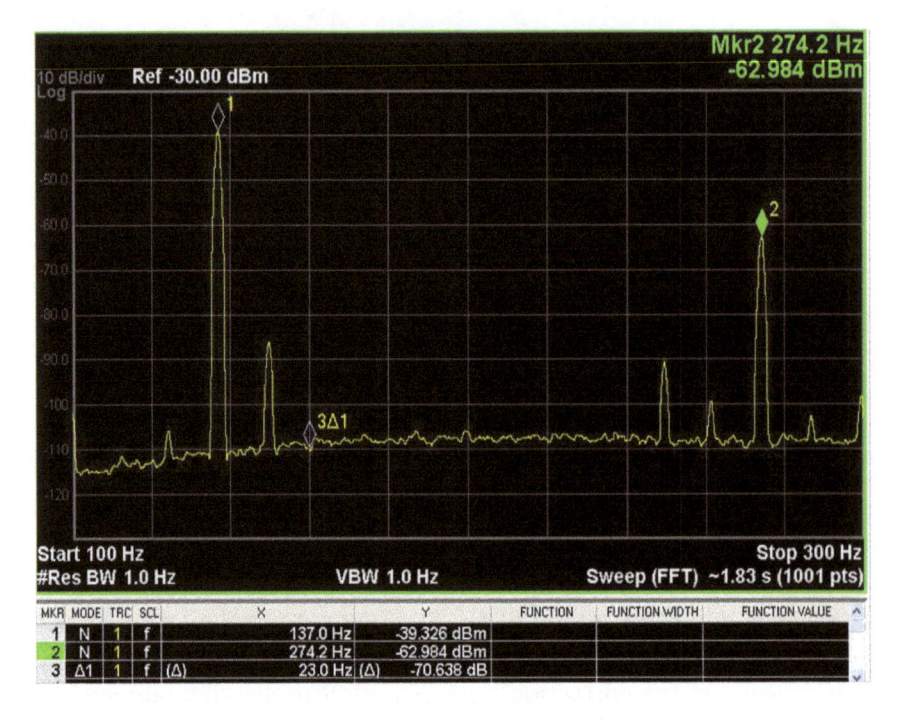

Fig. 7.14 Measured SNR of detected signal: 70.6 dB.

Table 7.1 Measured results of servo lock-in amplifier.

Parameters	Unit	Measured Results
DC Photocurrent	uA	10
Resonance line width	Hz	600
Gain	dB	108
Modulation depth	rad	3.5
Total Noise	pA/Hz	30
Min. Detectable Signal	dBm	−67
Dynamic range	dB	60
Sensitivity	V/Hz	1/60
Loop time constant	sec	1s

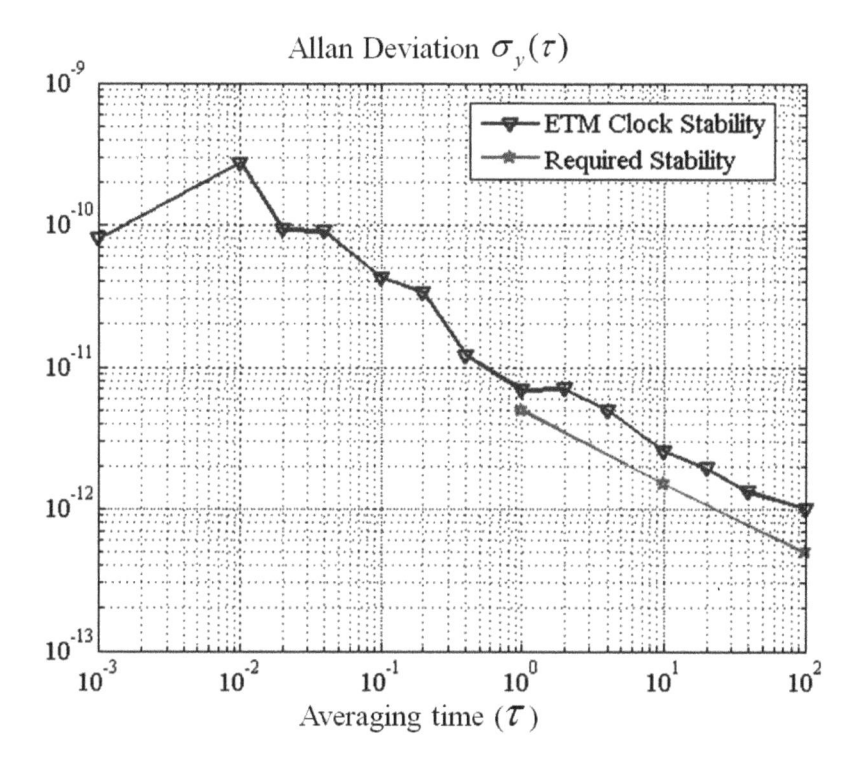

Fig. 7.15 Measured stability of the clock.

Summary

In this chapter is presented an analog phase-sensitive detection technique for the atomic spectroscopy applications in a low SNR environment. We report the development of a simple, high sensitivity, small sized, analog lock-in amplifier for the detection of very low-level signals. It has a dynamic range of 103 dB and the capability of recovering input signals in the nano-ampere range. The restrictions due to single supply operation require the redesign of the signal mixer, that effectively rejects the signal noise. In addition, the constraint of the power consumption in the space Rb clock, makes it necessary to use the low-power low-voltage electronic components, and a precise power management. The analog lock-in amplifier is designed according to these restrictions using space qualified components. The LIA functionality is verified using a low frequency constant current input source, to make it suitable for quantifying the amplitude and slope of atomic absorption signal i.e., the discriminator signal.

Figure 7.15 represents the realization of the goal of stabilizing the frequency of the OCXO/VCXO. The frequency stability is affected by the total gain, product of AC and DC amplifier gains, of the error signal and the time constant of the low-pass filter of the lock-in amplifier. The lock-in amplifier is able to produce an error signal that can keep the OCXO stabilized for a reasonable length of time. Under the right settings, the OCXO frequency can remain locked on the atomic transition for a much longer time. This chapter also describes the technique of simulation in TINA software for detecting low level optical signal, using the lock-in principle with trans-impedance amplifier. The TINA models contain all stages in detecting the optical signal, modulation, demodulation and filters. In order to simulate models for the real situation, a noise from the surrounding is applied in simulation. As a result, this model is used to tackle problems during developing hardware circuit. Experimental measurements confirm the capability of the developed locking system, to recover signals 60 dB below the noise level.

Chapter 8

Magnetic Shield Assembly for The Rb Physics Package

> The Magnetic shield is an integral and very important part of the Rb Physics package. The isolation of the components of the Physics package, from the environmental magnetic filed and its fluctuations, require special efforts in machining magnetic shield cyclinders of high permeability μ-metal. A magnetic shielding factor better that 10000 is required to get good clock stability. For that, the minimum of three layers of magnetic shields are considered necessary. In this chapter, a detailed study of magnetic shielding is included.

8.1. Introduction

The Rb atomic clock discussed in the book broadly meets the requirements of any satellite navigation system. However, the dimensions and weight of the clock can be reduced considerably, as the size of the microwave cavity is the only constraint. The Rb Physics package in the integrated filter cell configuration is developed, with three layers of magnetic shield. It has a magnetic shielding factor of 10^4. The Rb clock has the frequency stability of $5 \times 10^{-12} \tau^{-1/2}$ for a sample time up to 10^4 s.

We characterize the Rb discriminator signal, which demonstrates the line-width and SNR of the Physics package. We present experimental data, validating these two important features of Rb atomic resonance and their relation to the short-term frequency stability. In this chapter, the physical basis of the environmental sensitivity of the Rb frequency standard, particularly on the environmental magnetic field and its fluctuations, is discussed. In many of such state-of-the-art clocks, the environment sensitivity is the most significant performance limitation [226–227]. An understanding of their physical mechanisms is obvious concern for the development of RFS, especially, since it is intended for the space application. Thus, the experimental characterization, of all clock critical parameters and the principal factors that contribute to RFS instability, becomes necessary. The experiments show a number of design trade-offs, that can affect the clock performance. After verifying all design criticalities, it is possible to reach the stability $5 \times 10^{-12} \tau^{-12}$ and even 1×10^{-14} for the sample time $\tau \geq 10^4$ s. The simulation and the experimental work on the thermal and the vibration tests of the Physics package along with that relate to the electronics are described, for realizing the space qualified Rb atomic clocks. The thermo-vacuum model is designed to establish that it can be effectively used in the space clocks. The present work has laid the groundwork for the excellent satellite navigation clocks.

8.2. The Rb Physics package and its sensitivity factors

The core of the Physics package consists of the Rb bulb, integrated filter cell and the microwave cavity. The Physics package also includes a solenoid providing a constant magnetic field, a triple μ-metal magnetic shield layers to reduce the effect of environment magnetic field fluctuations and three bifilar heaters to control the temperature with uncertainty at the level of 5 milli-°Celsius.

8.2.1. Assembly of Rb bulb and lamp exciter circuit

The Rb lamp consists of electrode-less Rb bulb and a lamp exciter circuit. The bulb is filled with ^{87}Rb isotope and natural Rb of 99.99% purity in 1:1 ratio, and each in the quantity of 0.350 mg aprox. The use of the mixture of natural and ^{87}Rb in equal proportions ensures the minimization of the light shift [228–229]. The bulb is filled with Krypton/Xenon buffer gas at 2.0 ± 0.2 Torr, of nearly 99.995% purity, for ease of excitation as Krypton has low ionization potential.

Section 3.4 describes the details of the lamp exciter, which is excited by 100 MHz Clapp oscillator. The lamp is self RF heated,

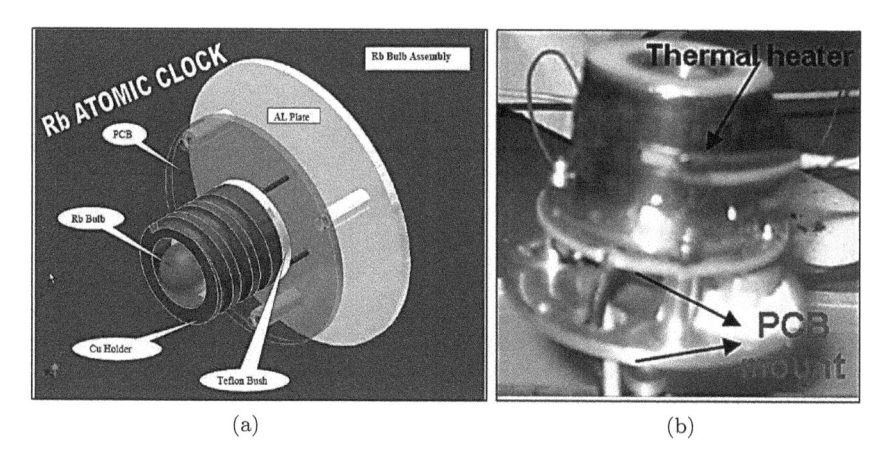

Fig. 8.1 The Rb bulb assembly Unit. (a) Simulated view (b) Experimental view.

but it requires additional temperature controller for stabilization of its intensity and its operation in the proper mode. The Rb bulb assembly consists of electrode-less Rb spherical bulb of dia 10 mm, surrounded by a coil, supported on Teflon bush and housed inside the copper cylinder. The copper cylinder is mounted on the FR-4 PCB. The PCB is attached to the aluminium. alloy mounting plate, shown in the Fig. 8.1. The Teflon bush works as an insulation between PCB and the copper cylinder. The Teflon bush has a replica, Fig. 8.1(a), of the Rb bulb for the proper fixing of the Rb bulb in the cavity. It assists in alignment of the bulb. The Rb Bulb assembly has to maintain 100°C. The external heaters are provided at the periphery of the copper cylinder with the necessary controller.

8.2.2. *Assembly of microwave cavity and photodiode detector PCB*

The TE_{111} microwave cavity tuned at 6.834 GHz is used for resonant transitions of the Rb atoms between the 0–0 hyperfine levels in the ground state. The absorption cell and the array of the photodiode detector are also kept inside the microwave cavity. To detect the hyperfine transitions and to obtain the clock signal, a number of photo-voltaic cells, with high efficiency in the near infrared (800 nm) region are mounted on a PCB. Photo detector PCB is kept at the

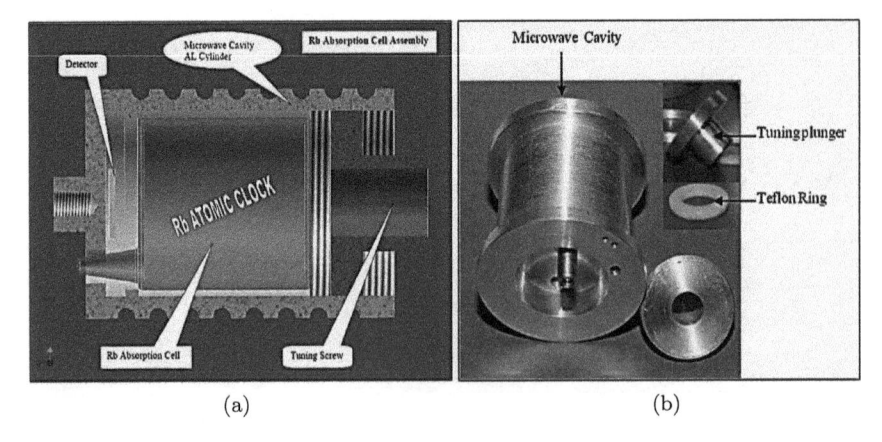

214   *Rubidium Atomic Clock: The Workhorse of Satellite Navigation*

Fig. 8.2 Assembly views of Microwave cavity. (a) Simulated view (b) Experimental view.

rear end of the Rb absorption cell. The back side of the PCB, on which these photodiode are mounted, is a ground plane which acts the termination of the microwave cavity. The Rb absorption cell has a Teflon ring over it to protect it from the damage inside the RF cavity. On the other end of the cavity, there is a tuning plunger with suitable threads for tuning the cavity precisely. The lock nut provision is kept to hold the tuning screw in the fixed position. The tuning plunger has an opening to transmit light signal. The output SMA connector is provided on the detector side, for the resonant microwave signal. The microwave cavity, with the Rb absorption cell inside, is provided with the bifilar heater winding, for maintaining the temperature at 75^0C with the temperature stability of better than 5 milli-^{0}Celsius. The simulated and real view of the microwave cavity are shown in Fig. 8.2.

8.3. Magnetic shield

Many instruments based on atomic spectroscopy, such as atomic clocks [230–232] or NMR gyroscopes [233–235] are sensitive to the magnetic fields. Stray magnetic fields and field fluctuations cause shifts of the atomic transition frequencies, which can lead to reduced stability. To achieve their full performance potential, it is necessary to shield such devices from the environmental magnetic fields and their fluctuations by enclosing them inside the shield of high-permeability material. That attenuates the ambient field and field fluctuations within the enclosure.

The field attenuation for large-scale magnetic shields is well documented in the literature. For the three-layer large scale shields of 10 cm diameter, a shielding factor of better than 10^7 is reported. For a constant wall thickness, the shielding factor for a high-permeability enclosure varies inversely to the size of the shield [236]. Therefore, the higher field attenuation may be expected for smaller diameter shields.

Here we discuss the design and testing of a set of three magnetic shields. The shielding factors are measured for the longitudinal and transverse directions of these cylindrical shields, both individually and in nested combinations. While the external magnetic field B_{ext} may be measured with a commercial Hall-effect Gauss meter. However, for the limited enclosed volume of the shields, and small internal fields, the measurement of the B_{int} is more challenging. We use two independent methods for measuring the internal magnetic field: a commercial miniature magneto-resistive sensor and a chip-scale atomic magnetometer [237–238]. The results from each method are in good agreement.

In Sec. 8.3.1, we discuss the requirement of the magnetic shielding in Rb clock. Section 8.3.2 describes the theory of magnetic shielding and formulae for estimating the shielding factor of specific shield design. The theoretical shielding factors of the three-layer nested shields are experimentally tested. The description of magnetic field control circuit is given in Sec. 8.3.3. The assembly of three magnetic shield with the Physics package are described in Sec. 8.3.4. In Sec. 8.3.5, we describe the required features of C-field current source.

8.3.1. *Requirements of magnetic shielding*

The magnetic field plays an essential role in the operation of Rb atomic frequency standards. The strict control of the magnetic field in the Rb Physics package is required for its normal operation. The satellite navigation Rb clock's accuracy goals, allow for a budget on the frequency magnetic sensitivity of the order of $\pm 1 \times 10^{-13}/G$. The magnetic field in the interrogation region is periodically measured, but between measurements, the field must not vary in such a way as to cause a fractional frequency shift larger than this value. In order to excite the transition between $F' = F+1, m_{f'} = 0$ to level F, $m_f = 0$, a constant dc magnetic field is needed in the Rb absorption cell and it must be parallel to the RF field which excites the transitions. The transition frequency between these two levels is

$$\nu = \nu_{hfs} + K_0 B_0^2. \qquad (8.1)$$

The value of K_0 is given by

$$K_0(\text{Rb}^{87}) = 575.14 \times 10^8 \text{HzT}^{-2}. \qquad (8.2)$$

From Eq. 8.1 the fractional frequency shift for a small fluctuation ΔB for an applied magnetic field of 2.5×10^{-5} T is,

$$\frac{\Delta \nu}{\nu} = 3.36 \times 10^{-4} \Delta B. \qquad (8.3)$$

If frequency stability is aimed better than 10^{-13} for the environmental or earth's field fluctuations by 10%, a shielding factor of at least 10,000 is needed.

8.3.2. Theory of magnetic shielding

The shielding factor S is defined as the ratio of the external field H_e, to the internal field H_i

$$S = H_e/H_i. \qquad (8.4)$$

In general, it is difficult to calculate the shielding factor for a high-permeability enclosure analytically, except for some simple cases. The reviews on the developments of shielding theory can be found in Refs. [236 and 239]. For a cylindrical shell, the shielding factor can be exactly expressed as:

$$S_t = \frac{1}{2}\mu t_1, \quad \text{(transverse field)} \tag{8.5a}$$

$$S_t = 2D\mu t_1 \left(1 + \frac{R}{L}\right)^{-1}. \quad \text{(longitudinal field)}. \tag{8.5b}$$

In these expressions, $t_1 \ll 1$, the relative permeability $\mu \gg 1$, is the ratio of the material thickness t to the shield radius R. L is the length of the shield and D is the demagnetization factor, which is a function of L/R. From Eq. (8.5), for L = 90 mm, t = 0.5 mm, R = 40 mm and $D = 0.75 (2 < L/R < 6)$, the calculated value of shielding factors are $S_t = 187$ and $S_l = 388$, which point out the limitation of a single layer shield. The high values of the static shielding factor are required. Therefore, it is necessary to go for the multiple layers of magnetic shield. In practice, we can exceed this limiting value by constructing multi-layered shields. In this case, the resulting shielding factor of a series of nested shields is proportional to the product of the individual shielding factors and some geometrical scaling factors, that approach unity for large inter-shield spacing. We design and test a set of three nested magnetic shields constructed of high-permeability material, with external surface areas for the individual shielding layers ranging from 157 to 960 cm^2. The transverse shielding factor of a system having three concentric cylinders is given by

$$S_t = \frac{1}{2}(\mu_1 \mu_2 \mu_3)(t'_1 t'_2 t'_3)(d'_{12} d'_{23} d'_{13}), \tag{8.6}$$

where the subscript i characterizes the ith shield and $d'_{i,i+1}$ represents the separation between shields i and i + 1. The thickness of cylinder walls are $t_1 = 0.5$ mm, $t_2 = 1$ mm, $t_3 = 1.2$ mm respectively. For three concentric cylinders, an approximate expression for S_l may be obtained by replacing the quantity μ_i by

$$4D_i\mu_i \times \left(1 + \frac{R_i}{L_i}\right)^{-1},$$

and putting

$$d'_{12} = \frac{1}{2}\left[1 - \left(\frac{R_2}{R_1}\right)^2\right], \quad d'_{23} = \frac{1}{2}\left[1 - \left(\frac{R_3}{R_2}\right)^2\right],$$

$$d'_{13} = \frac{1}{2}\left[1 - \left(\frac{R_3}{R_1}\right)^2\right],$$

and also replacing the quantity $d'_{i,i+1}$ by $f_i d'_{i,i+1}$ respectively in above equation of S_t. D_i is magnetization factor of shield cylinder i and f_i is a geometrical factor which nearly constant and has value 0.75 for $2 < L/R < 6$, and equal to about 0.9 for $L/R = 10$. In our case $D_1 = D_2 = D_3 = D$ and $f_1 = f_2 = f_3 = f = 0.75$ as $2 < \frac{L}{R} < 6$. So the longitudinal shielding factor for three cylinders is

$$S_l = \frac{64 S_t D^3 f^3}{\left\{1 + \left(\frac{R_1}{L_1}\right)\right\}\left\{1 + \left(\frac{R_2}{L_2}\right)\right\}\left\{1 + \left(\frac{R_3}{L_3}\right)\right\}}. \quad (8.7)$$

The value of demagnetization factor D depends on the ratio L/R [236].

$$D = \frac{\left(1 + \frac{R}{L}\right)}{2 \times f \left(1 + \frac{L}{R}\right)}. \quad (8.8)$$

For the different values of L/R, the values of the demagnetization factor D are shown in Fig. 8.3.

In the Rb Physics package discussed here, there are three concentric layers of μ-metal cylinders of lengths $L_1 = 145$ mm, $L_2 = 140$ mm, $L_3 = 91$ mm respectively and radii $R_1 = 51.7$ mm, $R_2 = 45.7$ mm, $R_3 = 45.5$ mm respectively. The average values are, $L = 125$ mm and $R = 47.6$ mm. We get from the Fig. 8.3 the demagnetization factor $D = 0.2542$. The transverse and longitudinal shielding factors are $S_t = 112860$, $S_l = 18599$ respectively.

8.3.3. *Magnetic field control circuit*

The shift in the second order Zeeman shift with the magnetic field variations for the F = 2, $m_f = 0$ to F = 1, $m_f = 0$ transition in the

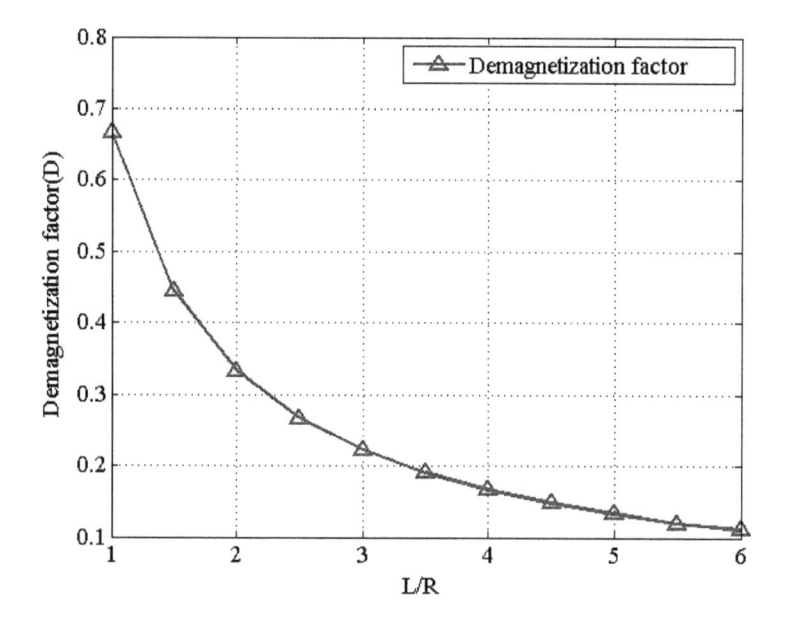

Fig. 8.3 The plot of demagnetization factor vs L/R.

Rb is,

$$\frac{\Delta \nu}{\nu} = 1.7 \times 10^{-3} B_0 \Delta B, \quad (8.9)$$

where B_0 is the C-field in mG and ΔB is the fluctuations in the field in mGauss. For a field B, equal to 244 mG, ΔB must be less than 40 mG. Therefore, in addition to shielding from the fluctuations in the magnetic field, a magnetic field control circuit must provide a highly stable C-field. The small constant magnetic field applied to the Rb atoms in the absorption cell, provides a quantum axis along the Rb light for the field independent F = 2, $m_F = 0$; F = 1, $m_F = 0$ transitions. It also provides some leverage for manipulating the hyperfine energy levels so that the hyperfine transition frequency could be exactly matched with the applied microwave frequency at the resonance. Thus, the uncertainty in the buffer gas pressure may also be taken care. The most important role of the applied DC magnetic field is to make Second's duration match with the SI unit of Time. The constant magnetic field of the order of 250–450 milli-gauss

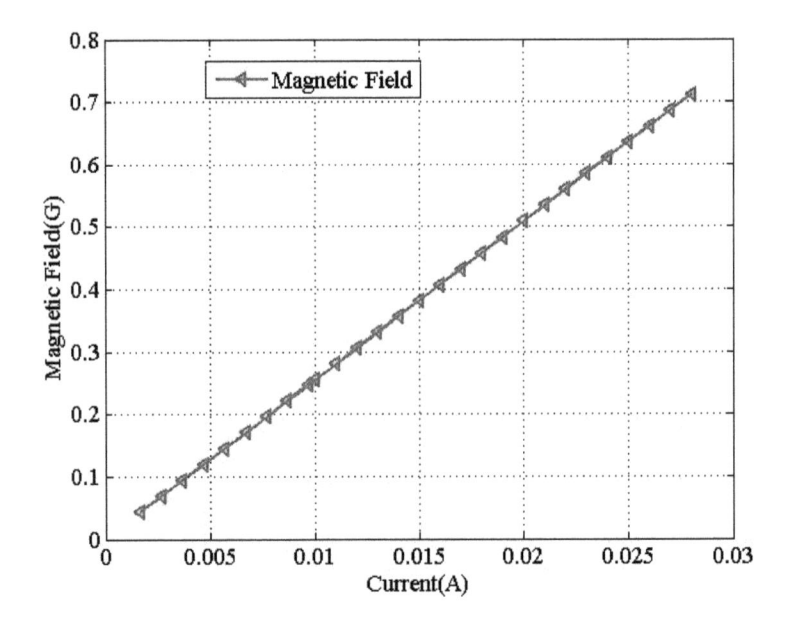

Fig. 8.4 Plot of the magnetic field variation with DC current.

is produced by a solenoid. The winding is done on a Hylem cylinder, a non-magnetic material. Its diameter is 70 mm and length is 120 mm. The winding is compact so that the homogeneity of the magnetic field is maintained. The microwave interrogation region lie well within the solenoid. The DC current versus the magnetic field plot is shown in Fig. 8.4.

8.3.4. *Assembly of magnetic shield*

The clock g.s field-free hyperfine transition frequency has the second order dependence on the magnetic field and is very sensitive to the fluctuations in the external magnetic field. Hence, in order to protect the Rb absorption cell and other parts of the Physics package from the magnetic field fluctuations, three concentric layers of magnetic shield, of high magnetic permeability, are used. The shield-1, contains the microwave interrogation region, with its end caps approximately 5 cm to the lamp assembly on the one side and 5 cm from the resonant cavity on the other side. The shield-2 covers

Table 8.1   Magnetic shield dimensions.

Shield	Length (mm)	Radius (mm)	Thickness (mm)
Inner shield	91	40.2	0.5
Middle shield	140	45.7	1
Outer shield	145	51.5	1.2

shield-1. The Physics package constrains, the spacing between shield layers to be small compared to their radii. Thus, the results of Gubser [240] apply, and we have a functional form to estimate the shielding effectiveness. Using Gubser's results and a finite element model, the shield mass and shielding effectiveness are optimized. The results of the optimization are shown in Table 8.1.

With a realistic and achievable mu-metal permeability, the model predicts the shielding factor, for this configuration, of order of 2×10^5. A prototype shield set with similar geometry for the ETM clock, has shielding factor of 2×10^4. All the three layers encircle the main cylinder assembly, as shown in Fig. 8.5(a). The magnetic shield layers are fastened to the main cylinder using 6-screws. The non-magnetic material (polyurethane) is used between layers. The end caps are provided to protect the Physics package from the external axial field and shown in the Fig. 8.5(b). The circular ends of the shield are fastened with the common cylinder, with a stud and the non-magnetic washers made of Teflon.

8.4. Base plate

In the space, the heat dissipation is a serious problem and requires very critical analysis. In order to achieve the temperature stability of the various components of the Physics package, it is important to keep thermal balance with the surroundings. This objective is achieved by placing the Physics package on a base plate for the heat dissipation. A base plate temperature controller (BTC) is provided to maintain the overall temperature sensitivity below the noise level. The base plate also plays an important role in reducing the effect of barometric fluctuations on the Physics package. The base

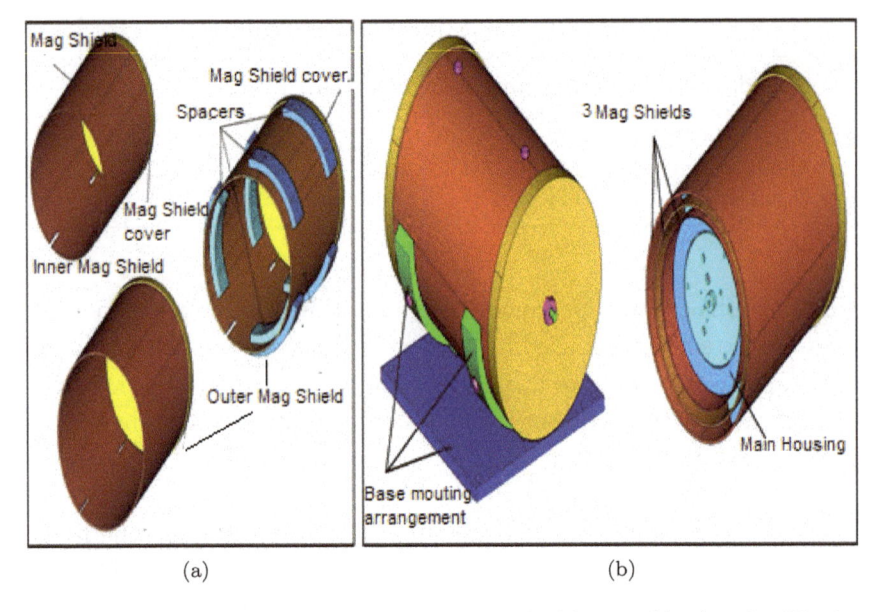

222   *Rubidium Atomic Clock: The Workhorse of Satellite Navigation*

Fig. 8.5 Schematic of three layer magnetic shield assembly for the Physics package.

plate temperature may be set with a precision temperature controller anywhere between 25 to 45°C. The dimensions of the Physics package also decide its thermal capacity. The simulation is done to get required information on the material, dimensions and temperature of the base plate.

Chapter 9

Integrated Testing and Characterization of Rb Clock Parameters

> The important task after developing Rb atomic clock is its characterization through stringent tests and evaluation of its performance. In this chapter, we discuss the details of testing, evaluation and characterization of the Rb atomic clock.

9.1. Introduction

The atomic correction signal of the Rb atomic clock is obtained by passing the light from a Rb lamp through an absorption cell, in the Rb Physics package. In the integrated filter cell technique, the absorption cell contains the natural Rb and Nitrogen as buffer gas at low pressure. The buffer gas reduces Doppler broadening in the microwave resonance [241]. In the absence of the microwave, the light intensity transmitted through the Rb vapour has the signature of the optical pumping [242–246], which removes atoms from the light absorbing hyperfine level, and population inversion is created. However, when the resonant microwave field is applied, the population inversion is disturbed and the transmitted light intensity decreases. Thus, the transmitted light intensity carries information on the Rb atom's response to the microwave field. This is the origin of the Rb atomic clock error signal. The OCXO frequency is locked on the position of the maximum optical absorption, by frequency modulating the 6.834 GHz resonant RF frequency, at modulating frequency fm = 137 Hz, and then demodulating the absorption signal from the photodiode, with the lock-in amplifier. The Rb atom's response to the modulated microwave signal is detected synchronously, which provides the correction signal having a dispersion-like shape. This correction signal is the heart of the clock functioning, and it provides an error voltage that ties the VCXO to the atomic resonance. Due to the non-linear interaction between the atoms and the microwaves, the atoms also produce a modulated signal at 2fm, the 2nd harmonic signal. The 2nd harmonic is a measure of the correction signal's quality, and consequently is a good indicator of the performance of the clock. However, the atomic dynamics of the 2nd harmonic signal's generation is not well understood. We discuss a series of studies on the 2nd harmonic signal and its dependence on the diverse clock parameters.

In the open-loop state, the output of the lock-in amplifier is monitored to get the information on the line-shape, transition widths, and frequency shifts. Then it is operated in a closed-loop mode in which, the voltage error signal is applied to the tuning port of the VCXO/OCXO. The corrected output of the VCXO is compared to a signal derived from a Hydrogen Maser. Figure 9.1(a) describes the

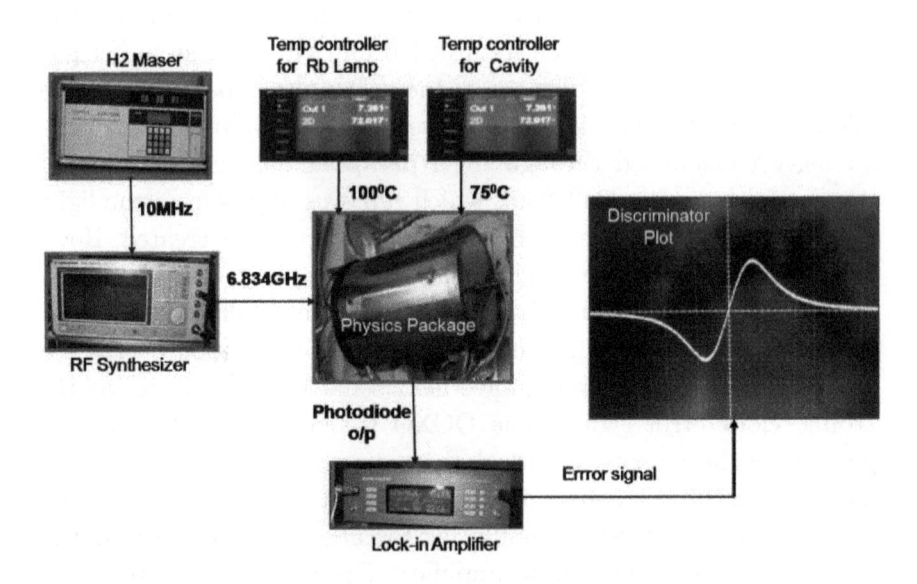

Fig. 9.1(a) The open loop experimental set-up of the Rb clock.

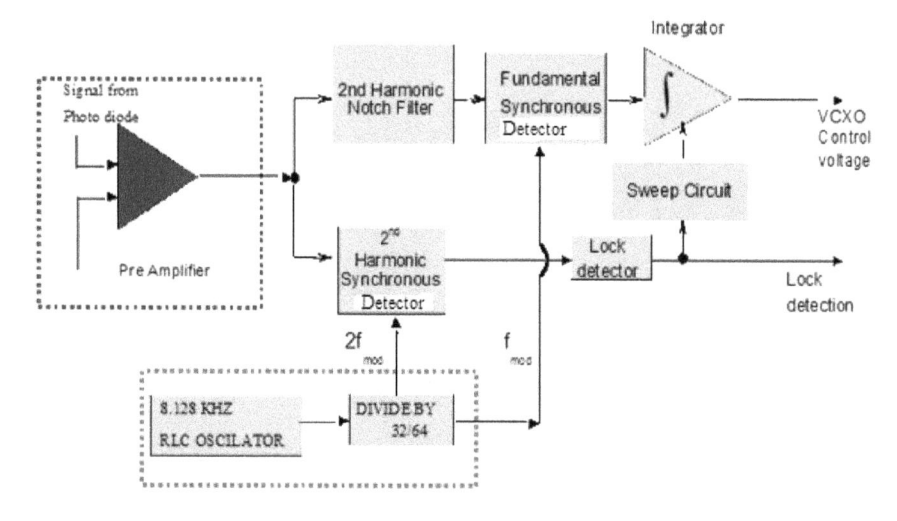

Fig. 9.1(b) Conceptual block schematic of Servo section.

open loop experimental set-up of the Rb clock. Figure 9.1(b) shows the servo-loop block diagram.

9.2. Critical clock parameters characterization

In this section, the open-loop characterization of the double resonances discriminator signal, a function of several experimental parameters such as, the absorption cell temperature T, modulation index and the microwave power $P\mu$ injected to the microwave cavity, is discussed. Figure 9.2 shows a typical experimental double-resonance absorption spectrum from the Physics package, as well as the corresponding discriminator error signal, obtained at the output of the lock-in amplifier.

The double resonance line-width and the absorption peak are obtained by plotting the resonance error signal with respect to the sweep resonance frequency in 8 KHz bandwidth. The resonance contrast is considered here as the ratio of the height of the error signal to the background DC voltage, at the output of the photodiode.

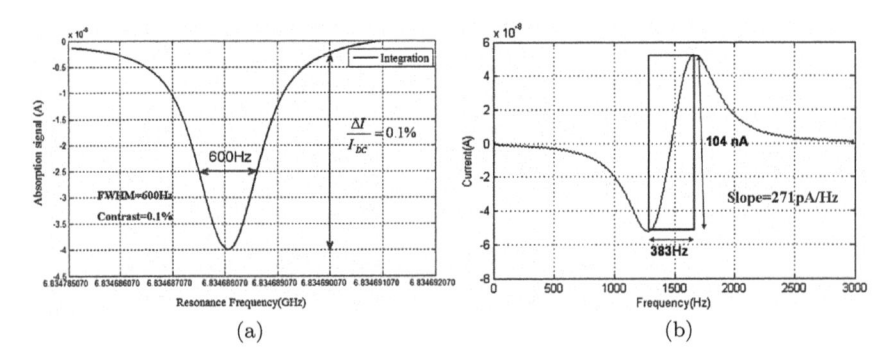

Fig. 9.2 (a) double resonance absorption spectrum and (b) slope of the discriminator error signal.

The absorption signal has a contrast of 0.1% and a narrow line width of 600 Hz, resulting in an error signal with an excellent discriminator slope of 271 pA/Hz.

9.2.1. *Resonance amplitude (2nd harmonic signal) vs. absorption cell temperature*

The amplitude of the resonance signal versus the absorption cell temperature is experimentally observed. The corresponding amplitude of resonance 2nd harmonic signal is plotted in Fig. 9.3. For the 25 mm dia absorption cell, we obtain a bell-shaped curve the 2nd harmonic signal amplitude is maximized in the temperature range 71.5°C to 72.5°C. The amplitude increases with temperature (T) up to a certain point (T of the order of 72°C), then drops rapidly at the higher temperatures. The signal is degraded by a factor 2 at 80°C compared to that at 72°C. It means that the clock short-term frequency stability is good at 72°C and degrades at 80°C. In the region of lower temperatures, the amplitude of the resonance is proportional to the density of the Rb atoms in the cell. Therefore, the temperature is optimized in a range 71.5°C to 72.5°C.

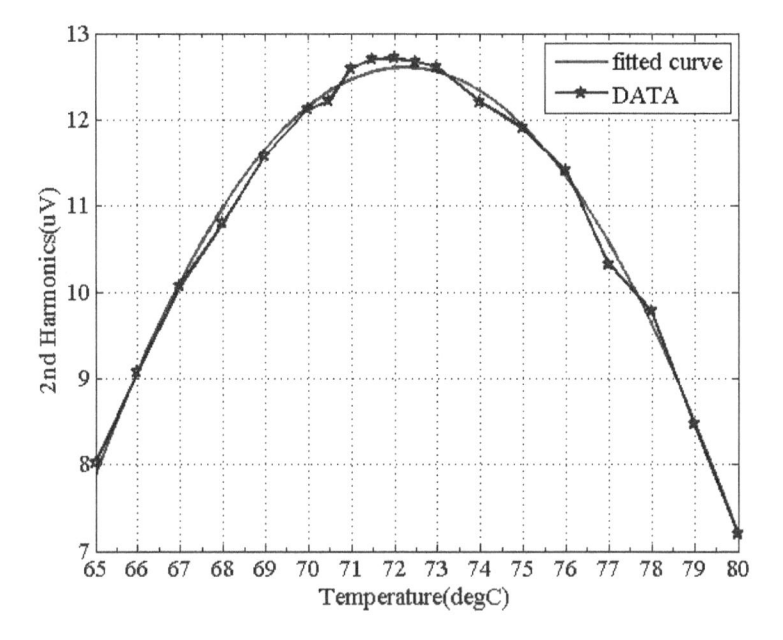

Fig. 9.3 The resonance 2nd harmonic signal vs. temperature plot.

9.2.2. Resonance (2nd harmonic) signal amplitude vs. RF power

To test the relation between the 2nd harmonic signal amplitude and the RF power, we perform the following experiment. The generalized 2nd harmonic signal is measured for different power levels of the microwave frequency. The power is varied between $-36\,\text{dBm}$ to $0\,\text{dBm}$ at $1\,\text{dBm}$ interval. Figure 9.4 shows the 2nd harmonic signal amplitude versus the RF power.

The 2nd harmonic signal increases linearly with the RF power, in the range of $-36\,\text{dBm}$ to $-19\,\text{dBm}$, and then it starts decreasing. We observe, that the noise in the detector increases strongly with increasing RF power above $-19\,\text{dBm}$. This is due to the increase of the shot noise associated to the background signal. The increase of

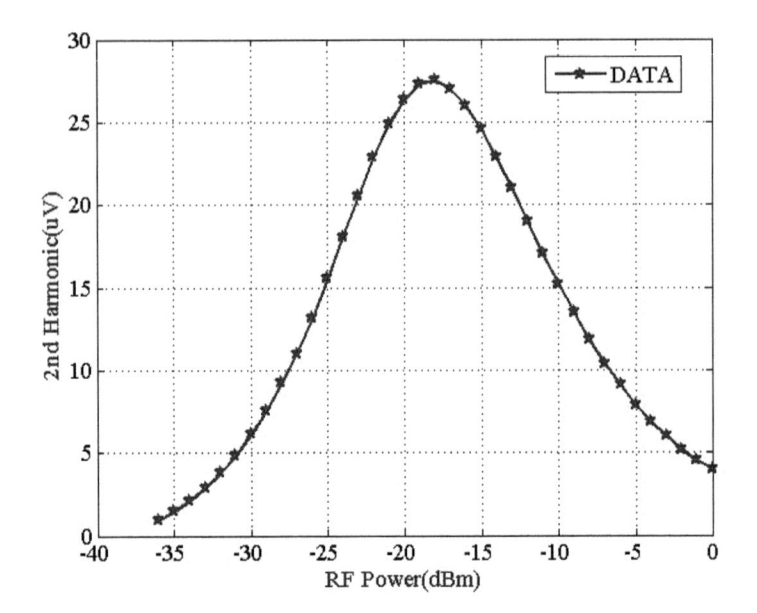

228   *Rubidium Atomic Clock: The Workhorse of Satellite Navigation*

Fig. 9.4   Plot of 2nd harmonic signal versus the RF power.

the AM noise in RF presents a 1/f slope at the modulation frequency. The increase in the RF FM noise occurs through a FM-AM conversion process in the vapour cell. Figure 9.5 shows that the RF power sensitivity of clock is 1×10^{-10}/dB. Therefore, to meet the clock's frequency stability specification, we need RF power stabilization within the range of ±0.05 dB.

9.2.3. *Resonance (2nd harmonic) signal amplitude and line-width vs modulation index*

The typical experimental plot of the amplitude and the line-width of the 2nd harmonic signal versus modulation index is shown in Figs. 9.6(a) and 9.6(b) respectively.

These results are for a cell of dimensions, 25 mm long and 25 mm diameter. The cell is filled with 10 Torr of a mixture of Ar/N$_2$ in the partial pressure ratio of ~1.4, at a temperature of 75°C. As

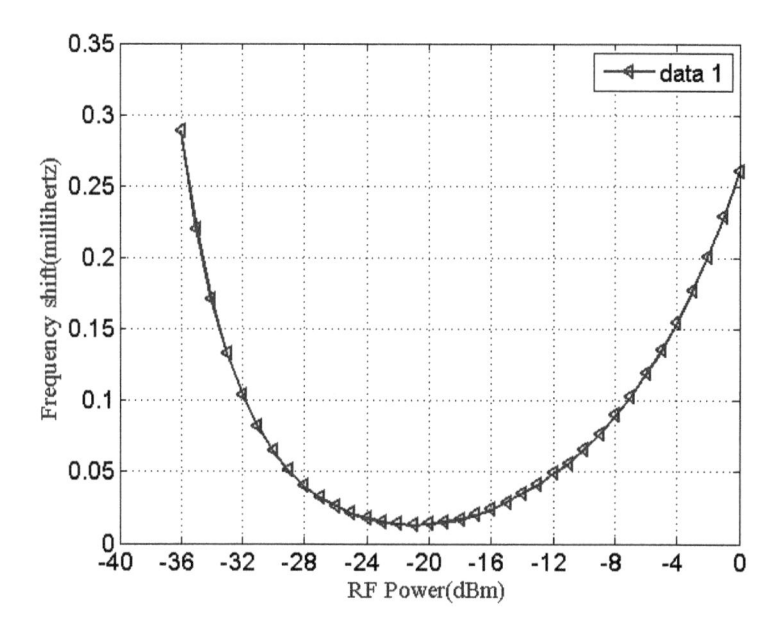

Fig. 9.5 The variation of clock frequency with RF power.

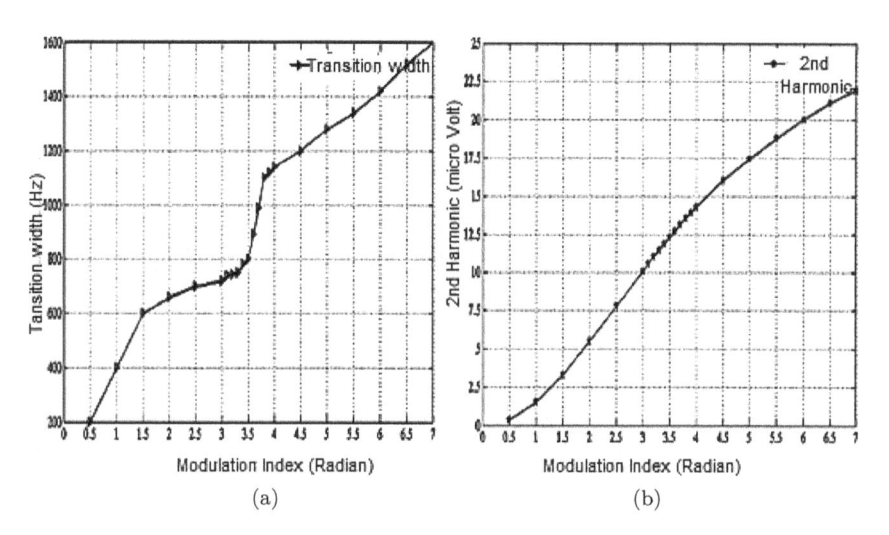

(a)

(b)

Fig. 9.6 Plot of (a) Transition width vs. modulation index and (b) 2nd Harmonic signal vs. modulation index.

it is readily observed, the 2nd harmonic signal and resonance linewidth are directly proportional to the modulation index. It provides a means of direct optimization of the modulation index. In the present situation, the 2nd harmonic signal appears to increase to the maximum value of 22uV at a modulation index of 7 radian and the transition line-width is 1.6 kHz. As expected, for the modulation index greater than 3.5 rad, the resonance signal increases and the absorption signal line-width gets larger. Therefore, measurements are done with different modulation indices and the results for the ratio of 2nd harmonic amplitude to resonance width are shown in Fig. 9.7.

From the above plot, it is clear that the at the modulation index of 3.5 radian, the ratio of the 2nd harmonic signal to transition linewidth is the maximum. Therefore, the modulation index of 3.5 rad is fixed for phase modulator.

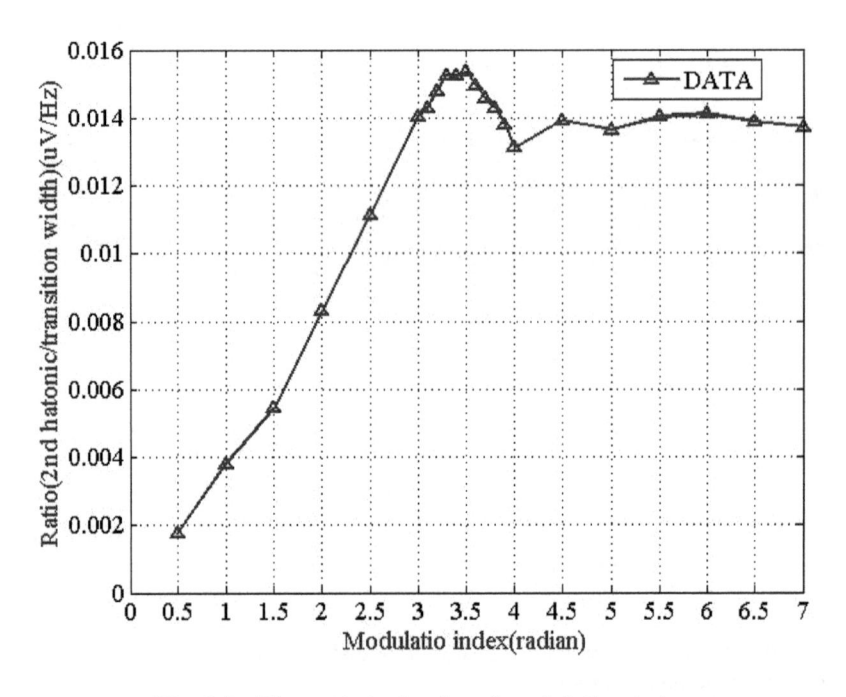

Fig. 9.7 The optimized value of modulation index.

9.3. Analysis of Rb resonance frequency dependence on various parameters

9.3.1. Temperature sensitivity

The resonance frequency of the Rb, in the absorption cell, depends on the temperature and the nature of the buffer gas, and is thus sensitive to the fluctuations in the environmental temperature. This fact is generally reflected in the long-term frequency fluctuations. It is a standard practice to use a mixture of two different buffer gases with opposite temperature coefficients. In the case of separate filter cell configuration, Nitrogen with a temperature coefficient δ of $+0.52$ Hz/°C/torr and Argon with $\delta = -0.32$Hz/°C/Torr are of particular interest [247–248]. In this configuration only ^{87}Rb isotope is used along with buffer gases.

A mixture of buffer gases with an appropriate pressure ratio can provide in principle, a zero temperature coefficient. However, in these experiments with integrated filter cell configuration, a cylindrical absorption cell of 25 mm of diameter and length, containing natural Rb and 9.5 torr of Nitrogen, is used in the filling process, and the temperature coefficient of the resonance frequency of the double resonance line for at 6.834 GHz is measured. To control the temperature of the Rb lamp and the integrated Rb absorption cell, separate heaters are provided. In addition to these two heaters, one common heater for maintaining the temperature a few degrees above the ambient temperature is provided. By using the double oven with the temperature controllers, the temperature stability of 5 milli Celsius is achieved. The temperature of the absorption cell is changed from 55°C to 95°C. The observed fractional frequency offset of 10 MHz output versus the temperature is shown in Fig. 9.8. A residual temperature coefficient of the order of 7×10^{-11} is measured. This is due to the presence of a quadratic coefficient in the dependence of the hyperfine frequency on the temperature. It is possible to adjust the actual mixture ratio in order to match the condition of zero temperature coefficient at the desired temperature of operation [249].

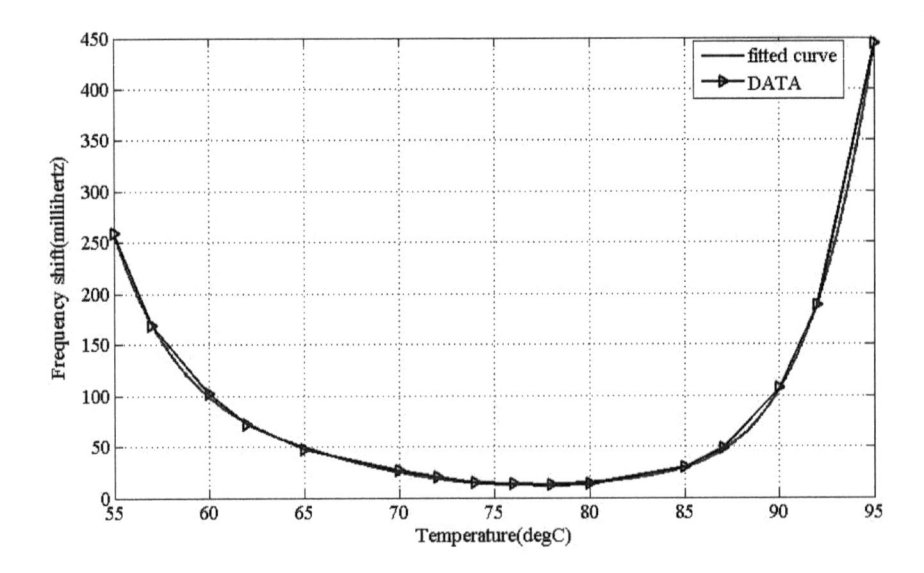

232   *Rubidium Atomic Clock: The Workhorse of Satellite Navigation*

Fig. 9.8   Absorption cell temperature vs Rb resonance frequency.

9.3.2. *RF power sensitivity*

The Rb atomic clock's RF power sensitivity is primarily due to its inhomogeneity within the absorption cell, as the microwave field strength is not uniform inside the cavity. The variations in the microwave power at different locations within the cavity produce a spatial dependence of the interrogated atoms. Any resonance line spatial inhomogeneity or asymmetry results in the locked frequency's variation with the RF power. If other frequency determining variables, such as C-field, temperature, light intensity, or light profile are stable, then a change in the RF power causes a frequency change. The RF sensitivity is experimentally tested, and plotted as shown in Fig. 9.5. For the Rb clock, RF power coefficient is about 1×10^{-10}/dB. The requirement of the long term frequency stability of 1×10^{-12} imposes a stringent control of ± 0.01 dB on the RF power. It is thus, established that the RF power instability is also a contributor to the long-term instability of the Rb clock.

9.3.3. Magnetic field sensitivity

A DC magnetic field is required to provide an axis of quantization to the system for the fulfilment of the selection rule and gives a certain freedom by which the atomic resonance frequency can be fine-tuned to the desired RF synthesizer frequency. The magnetic field also removes Zeeman degeneracy in the ground state. The $m_F = 0$ hyperfine sublevels are shifted quadratically, causing a shift of the hyperfine frequency according to the equation [250]:

$$\Delta \nu_B = 575.14 \times 10^8 B_0^2 \text{Hz}, \tag{9.1}$$

where B_o is the magnetic induction in tesla. This field is created by means of a solenoid inside a magnetic shield. It is of the order of a few tens of μT and produces a shift of the order of 10 Hz. The current which produces this field and the shielding factor of the magnetic shielding layers must be compatible with the frequency stability requirement of the Rb atomic clock.

9.3.4. SNR vs. short term stability

The contrast ratio (C $\approx$ 0.1%) is very small in case of double resonance based Rb clock, and it is very difficult to detect experimentally. Therefore, the short-term frequency stability characterization of the passive Rb frequency standard is done by calculating the signal-to-noise ratio. The measurements are made on the experimental setup shown in Fig. 7.13. The time-domain frequency stability of the passive RAFS is expressed as follows [249–250]:

$$\sigma(\tau) = \frac{k\tau^{-1/2}}{\left(\frac{\nu_{hf}}{\Delta \nu}\right) \times SNR}, \tag{9.2}$$

where ν_{hf} is the atomic reference frequency, $\Delta \nu$ the full width of the resonance, SNR is the signal-to-noise ratio of the Physics package, k is a constant with value 0.1$\sim$0.2, τ is the sample average time. We make measurements of the signal-to-noise ratio the output from the lock-in amplifier. The output SNR is measured with a spectrum analyser, Fig. 7.14, as a function of the frequency and it about 70 dB.

9.3.5. Frequency stability of the Rb atomic clocks

The Rb atomic clock frequency stability can be measured by comparing its 10 MHz output to that from H-Maser or Cs atomic clock, using high precision frequency counter or phase recorder. The Allan deviation or two sample stability measurements for different time intervals or sample times can be computed. The double mixer technique is also very accurate means of determining the frequency stability. The results give the signal level at 137 Hz in a 1-Hz bandwidth. It is observed that with the atomic line Q of 2×10^7, and a signal-to-noise ratio of 70 dB, the clock's frequency stability is of the order of 3×10^{-12} for an averaging time of 1s.

The frequency shift measurements are performed by calculating the offset between the measured F = 2, m_F = 0; F = 1, m_F = 0 of ^{87}Rb resonance frequency and the unperturbed ^{87}Rb atom frequency, 6.834682612 GHz. The ETM (engineering thermal model) of Rb atomic clock is developed with the optimized parameters such as modulation frequency and modulation index, RF power of local oscillator, temperature of the absorption cell and Rb bulb, magnetic field and finally the time constant of loop filter. The tested ETM model is shown in Fig. 9.9. The space Rb atomic clocks have medium term accuracy of the order of a few parts in 10^{-13} or 10^{-14}. The closed loop is stabilized on the zero crossing point of the discriminator signal. In the locked state, the error signal is fed to EFC of a 10 MHz OCXO through a PI controller. This OCXO serves as the reference to the microwave synthesized source for the microwave cavity and it also provides the stabilized 10 MHz output of the clock.

With the all sub-systems in the normal operation, the short-term frequency stability of the locked VCXO or the Rb atomic clock is $6.85 \times 10^{-12}/\sqrt{\tau}$ for the integration time in the range of 1s $\leq \tau \leq$ 100s, as plotted in the Fig. 9.10. The best observed Allan deviation value is 1×10^{-12} at 100 sec. It is also in good agreement with the theoretical stability limit of $\approx 3.3 \times 10^{-12}\, \tau^{-1/2}$, calculated from the Eq. (9.2). Figure 9.10 also shows the frequency stability plot of the free running OCXO and locked OCXO.

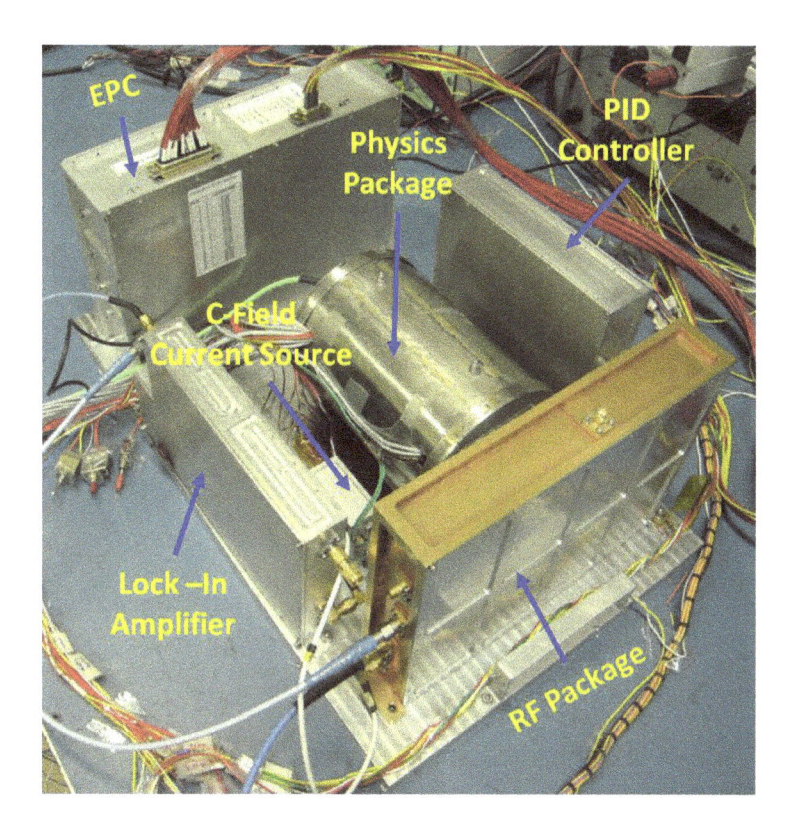

Fig. 9.9 Open view of tested ETM of the Rb atomic frequency standard.

9.4. The outer space and radiation effects

The radiation is a major consideration in the design of the space Rb atomic clocks. In outer space, the Rb atomic clock is continuously exposed to energetic particles and secondary radiation. Not only is the VCXO affected by total dose and bursts, but the servo electronics in the Rb atomic clock is also subjected to transitory loss of lock due to the radiation. Some key components in the Rb atomic clock are radiation hardened to significantly reduce the impact of radiation. Heavy radiation shielding is also added to minimize the impact of the space radiation. In spite of all this, during periods of intense solar activity, the Rb atomic clocks have clearly exhibited discernible sensitivity to the radiation. The space Rb atomic

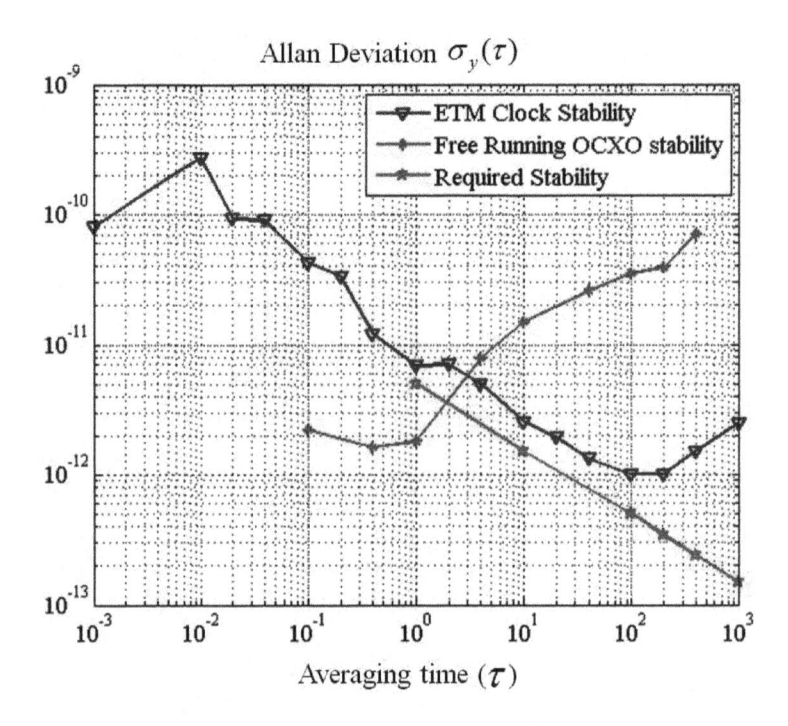

Fig. 9.10 Allan deviation plot of unlocked and locked OCXO.

clocks must maintain a high level of accuracy and frequency stability for several years, under the harsh environment of space. The mechanical design must be such that the Rb atomic clock withstands the enormous shocks and vibrations at the launching. The extreme temperatures that may be encountered in space, require that the thermal and mechanical designs be such that the Rb atomic clock maintains excellent frequency stability over a broad range of temperatures. This requirement is expressed in terms of the Temperature Coefficient (TC) of the Rb atomic clock. Typically, the magnitude of change in output frequency of a Rb atomic clock should be of the order of $5 \times 10^{-13}/°C$ for the temperature variations between $-10°C$ and $+50°C$. This performance should be achievable in the high vacuum of the order of 10^{-5} Torr in the space. Another stringent requirement for the space Rb clocks is the electromagnetic interference (EMI). EMI gaskets, feed-through filter capacitors, filter boxes and semi-rigid shielded coaxial cables are used to reduce the EMI effects. Another important space requirement is the magnetic sensitivity. Typically, the magnetic coefficient should be no greater than 10^{-12}/Gauss. Several layers of mu-metal shields are used to meet this requirement.

Summary

We have discussed ETM of space qualified Rb frequency standard. These devices have dimensions, the power dissipation depending of the degree of miniaturization. The microwave cavity mode and its type can influence the physical dimensions of the Rb atomic clock. Among all the atomic clocks, the Rb atomic clock is the only one which can be highly miniaturized in size and weight. This is the reason of its preference to other atomic clocks, for the satellite navigation. It has the short-term stability better than $6.85 \times 10^{-12} \tau^{-1/2}$. The influences of the three dielectric materials (absorption glass cell, Teflon ring, FR4 photodiode PCB) on the resonance frequency and cavity Q are analysed. The level of EM field, an important parameter in the microwave cavity design, is optimized and experimentally verified. The experimental cavity measurements are compared with the analytical results and have a close agreement to simulated values.

The development phase of the ETM model for satellite navigation program is important for achieving space qualification of the Rb atomic clocks. The critical design and manufacture readiness are equally important. A qualified model of Rb clock should undergo stringent qualifying testing. After the qualification and production go-ahead, the space qualified Rb atomic clocks for satellite navigation may be commercially produced. In the next chapter, the recipe for the Rb atomic clock is discussed.

Chapter 10

Recipe of Rb Atomic Clock

> Here, we present the complete details on how to develop and also manufacture Rb atomic clocks. Many companies, in the business of frequency and time standards and allied measuring equipment, may find the recipe very handy. This is perhaps, for the first time that exhaustive details on the Rb atomic clocks are provided in one place. One need not shuffle large numbers of papers and literature on Rb atomic clocks for its development.

10.1. Introduction

In the previous chapters, we have described various components of the Rb atomic clocks, in details, giving perspective of Rb atomic clock. We freely, use the information given in various chapters, to explicitly, develop the Rb atomic clocks. In this book, at many places, the dimensions, parameters and characteristics of the Rb atomic clock are mentioned. These values are indicative only, for those who want to develop the Rb atomic clocks first time. As a matter of fact, highly miniaturized Rb atomic clocks can be developed and indeed have been developed. But the miniaturization should not be a hindrance for developing the Rb atomic clock for the first time. To begin with, we first describe ultra-high vacuum (UHV better than 10^{-6} Torr) system for filling the Rb bulbs and Rb absorption cells. The glass technology is very crucial and important, as the choice of material of the glass and making moulds for the Rb bulbs and the absorption cells, all come under this critical technology.

10.2. Rb Physics package

10.2.1. *Rb bulb and absorption cell filling system*

The block diagram and photograph of UHV system are given in Figs. 10.1 and 10.2. The UHV unit consists an integrated turbomolecular pumping system, which can give vacuum of the order of 10^{-8} Torr. The RbCl and Calcium (Ca), both of the highest purity 99.999%, are placed inside a quartz ampule. This ampule is fused to Pyrex 7070 /Schott 8436/GE-180/C48-3 bulb/cell manifold, using graded seal, and is connected to the all metal vacuum system through the bake-able vitreous O-ring as shown in Fig. 10.1. To clean and remove moisture, the vacuum system is baked continuously for several hours, at the temperature between 350–400°C. During the baking, the vacuum system is kept running so that the impure gases are flushed out. After three to four cycles of baking, the vacuum system is flushed in with high purity Nitrogen gas. It needs to be mentioned here, that the glass manifold bulbs/cells are thoroughly cleaned with hot chromic acid and rinsed several times with triple distilled and de-ionized water, before these are joined to UHV system. The glass manifold and bulbs/cells also are baked several times, for long hours, to drive out the moisture completely. The vacuum system with the attached glass manifold, is run again to achieve the vacuum better than 10^{-6} Torr. The specifications of the Rb cells and bulbs for IFT are mentioned in Sec. 10.2.2 and the dimensions are shown in Figs. 10.3 and 10.4 respectively. The dimensions are only indicative and may vary, if dielectric material is used to fill the space between the walls of microwave cavity and that of the absorption cell. In SFT a separate ^{85}Rb filter cell is inserted between the Rb bulb and the absorption cell. In SFT, isotopic Rb is used in all three glass components namely the Rb bulb, filter and absorption cells. The filling specifications for SFT are mentioned in Sec. 10.2.3.

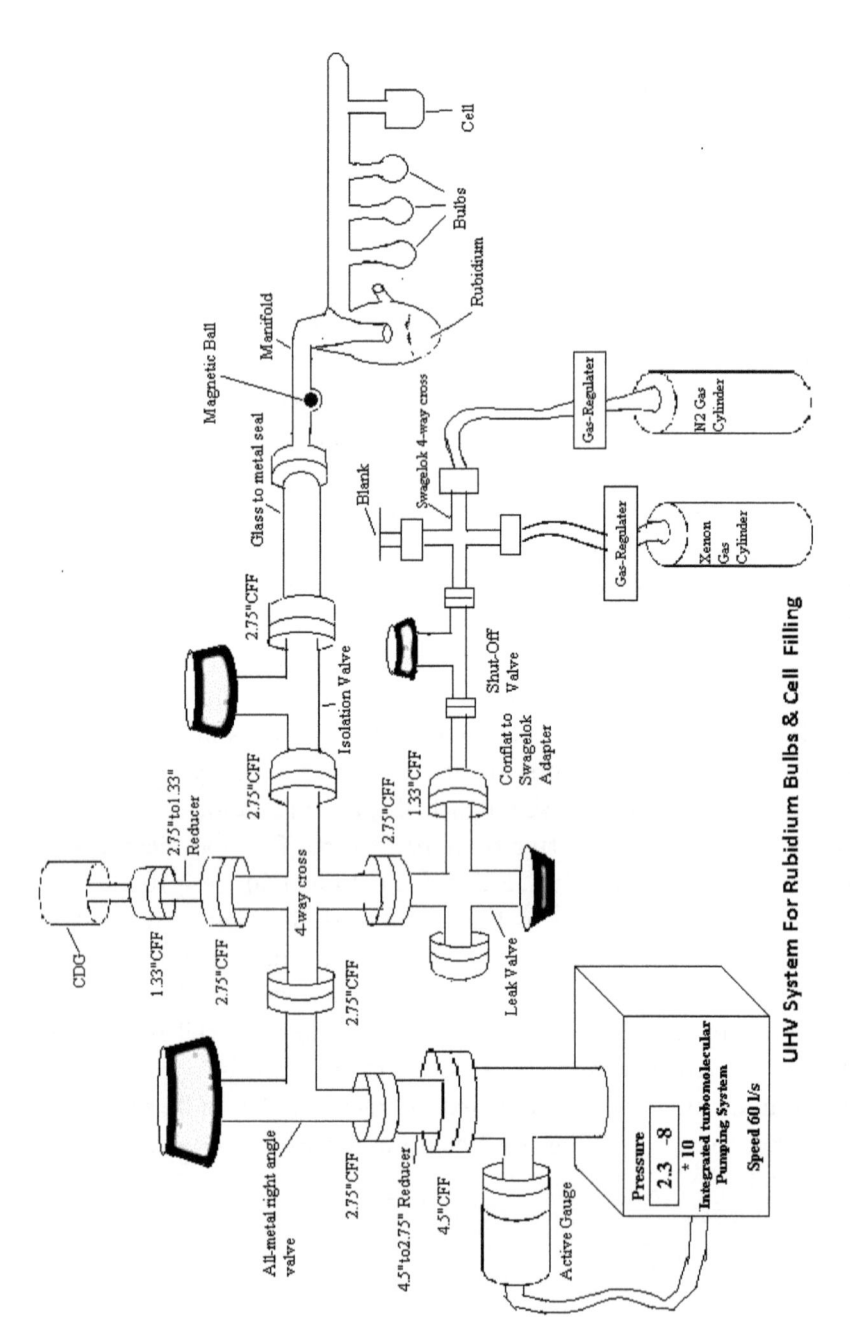

240   Rubidium Atomic Clock: The Workhorse of Satellite Navigation

Fig. 10.1   Diagram of Rb bulb and cell filling ultra high vacuum system.

Recipe of Rb Atomic Clock 241

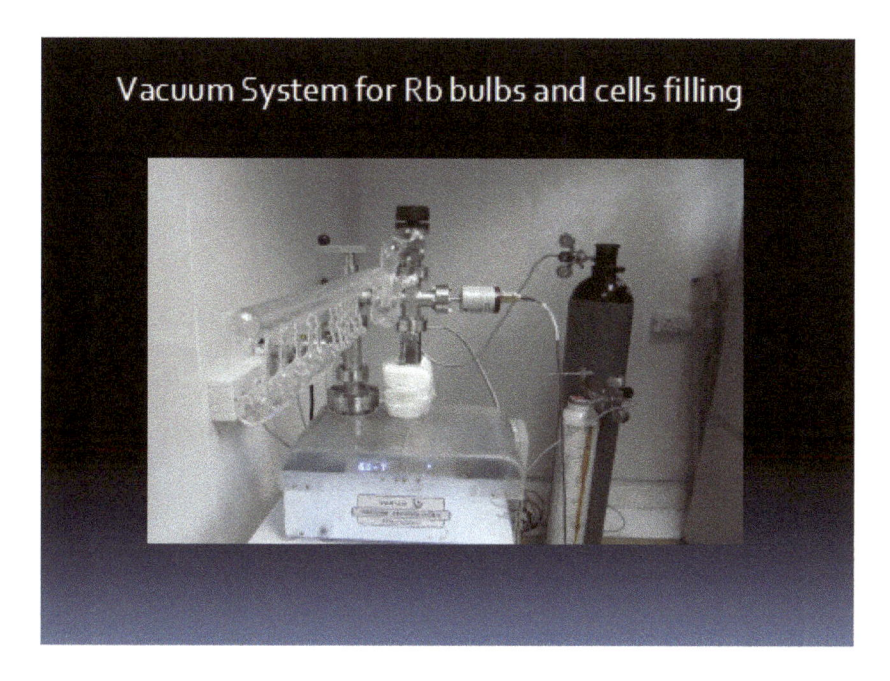

Fig. 10.2 Rb bulb/Cell filling ultra- high vacuum system.

10.2.2. *For integrated filter cell technique*

10.2.2.1. *Technical specification of the Rb absorption cells*

1. Cells must be made of C48-3/Schott 8436/GE-180 glass as shown in the drawing (Fig. 10.3).
2. Cells must be filled with 1.5 ± 0.1 mg of natural Rb.
3. All cells must be filled with Nitrogen gas at 9.5 ± 0.5 Torr of 99.999% purity.
4. The proper filling of the cells and their quality should be provided with the standard technological process and measuring equipment.
5. The quality control also includes the check of the external view and dimensions in accordance with the drawing (Fig. 10.3).

10.2.2.2. *Technical specification of the Rb bulbs*

1. Bulbs must be made of C48-3/GE-180 glass as shown in the drawing given in Fig. 10.4.

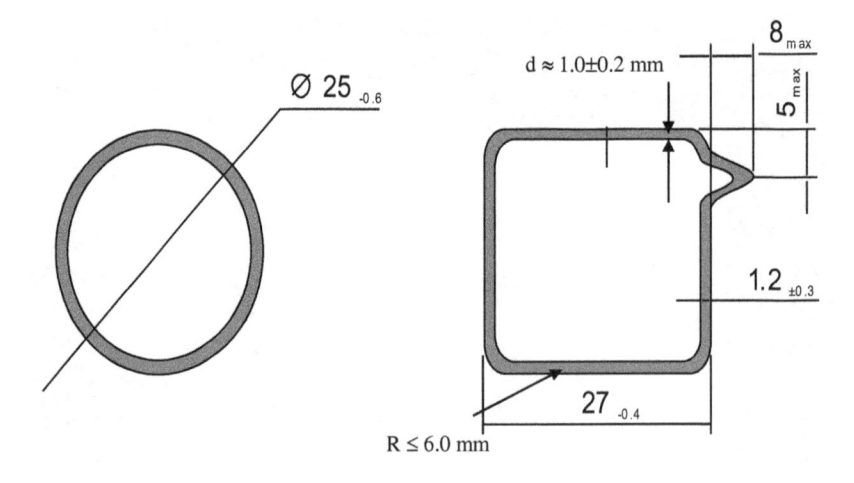

Fig. 10.3 Drawing of the Rb absorption cell.

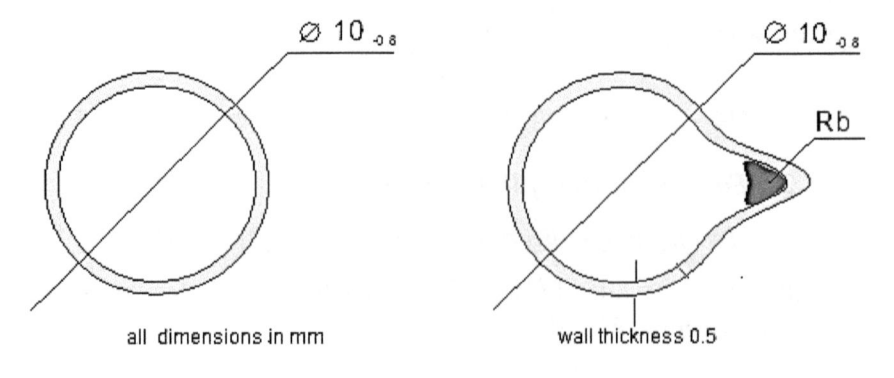

Fig. 10.4 Drawing of Rb bulbs.

2. Bulbs must be filled with ^{87}Rb isotope of 99.99% purity and natural Rb of same purity, in the 1:1 ratio and each in the quantity of 0.350 ± 0.1 mg.
3. Bulbs must be filled with Krypton or Xenon gas at 2.0 ± 0.2 Torr of 99.995% purity.
4. The proper filling of the bulbs and their quality should be provided with the standard technological process and measuring equipment.
5. The quality control also includes the check of the external view and dimensions in accordance with the drawing (Fig. 10.4).

10.2.3. Specifications for separate filter cell technique

10.2.3.1. Technical specification of Rb bulbs

1. Bulbs must be made in C48-3/GE-180 glass o.d 10mm wall, thickness 0.5–0.7mm.
2. Bulbs must be filled with ^{87}Rb isotope of 99.99% in the quantity of 0.5+0.1 mg. Bulbs must be filled with Krypton/Xenon gas at 2.0 ± 0.2 Torr of 99.995% purity.

10.2.3.2. Technical specification of Rb filter cells

1. The Filter cells must be made of C48-3/Schott 8436/GE-180 glass o.d=33 mm, L=15mm wall thickness 1 ± 0.2mm. Sealing notch 5mm from the edge and on one of the circular discs with sealing tube length as small as possible, and must be filled with 1.0 ± 0.1 mg of ^{85}Rb isotope of 99.99%.
2. All filter cells must be filled with Nitrogen gas at 50 ± 0.5 Torr of 99.999% purity.
3. The proper filling of the filter cells and their quality should be provided with the standard technological process and measuring equipment.
4. The quality control also includes the check of the external view and dimensions as per specs.

10.2.3.3. Technical specifications of the Rb absorption cells

1. Cells must be made of C48-3/Schott 8436/GE-180 glass O.D=27mm, L =26.5mm Cell wall thickness 1.0 ± 0.2 mm. Sealing notch 5 mm from the edge and on one of the circular discs with sealing tube length 7 to 10 mm and tube's o.d=5 mm.
2. It must be filled with 1.0 ± 0.1 mg of ^{87}Rb isotope of 99.99% purity.
3. All cells must be filled with Argon and Nitrogen gases with ratio (Ar/N2)=1.538 at 21.7 ± 0.5 Torr and of purity 99.995%.
4. The proper filling of the cells and their quality should be assessed with the technological process and measuring equipment.
5. The quality control also includes the check of the external view and dimensions as per specs.

10.2.4. Rb bulb and absorption cell filling process

For filling the bulbs/cells, the ampule containing the Rb chloride and calcium is heated gradually. As it is heated, there is out-gassing and the pressure rises. The heating is interrupted several times so that vacuum is better than a certain cut-off level ($\sim 10^{-4}$ Torr). This cycle of heating is repeated till out-gassing stops and vacuum stabilizes in UHV region. Now, the UHV valve is closed to isolate glass manifold and the ampule is heated gradually to 600 to 700°C. At this temperature, the chemical reaction starts and Rb in silvery vapour form appears. The following chemical process takes place:

$$2\,\text{RbCl} + \text{Ca} = 2\text{Rb} + \text{CaCl}_2.$$

The Rb vapour is driven, using a fine-nozzle flame, to the bulbs or cells, whichever is to be filled. The bottoms of the bulbs/cells are cooled using a copper tube, through which chilled water flows. The copper tube should touch bottoms of bulbs or cells. The area touched should be same for all bulbs, or cells, whichever are being filled. This helps in condensing the Rb vapour at the bottoms of bulbs/cells in almost equal quantity. The experience helps in assessing the quantity of the Rb deposited as a film in the bulbs/cells. The heating is stopped now and the UHV valve is slowly opened to achieve UHV. Once it is achieved again, the UHV valve is closed and the needle diaphragm valve is opened to admit ultra-pure gas/es in the bulbs/cells, in controlled manner till the desired gas pressure of a few Torr is achieved. Now, the needle valve is closed and the sealing process is initiated. There are two constrictions in tube connecting bulbs/cells to the manifold. At first, the sealing is done at the constriction farthest from the bulbs/cells, to ensure pressure in the bulbs/cells does not rise, and then at the constriction nearest to the bulbs/cells is sealed. The bubs/cells are detached from the manifold. This way the Rb bulbs/cells are filled with natural/isotope Rb and the buffer gas/es at the required pressure. It is also possible to introduce combination of buffer gases by attaching ultra-high pure gas cylinders through needle/diaphragm valves. The gas pressures are measured by the capacitance gauges (0–100 torr).

The above procedure for producing Rb bulbs/cells is the most critical part of developing the Rb atomic clocks. The filling process decides the quality of the Rb bulbs/cells, which finally determines the life of the Rb atomic clock. The lives of the Rb atomic clocks depend mainly on the proper functioning of Rb bulbs/cells. Once the bulbs/cells are depleted of Rb metal, the Rb atomic clock fails. So, it is necessary that the utmost care is taken while filling these Rb bulbs/cells. If the Rb bulb/cell filling is good, then these may last 10 years or more. We now discuss, the main components of the Physics package, with emphasis on their development.

10.2.4.1. Rb lamp

The Rb lamp consists of electrode-less Rb bulb and a lamp exciter. The Rb bulb is made of C 48-3/ Schott 8436/GE-180 glass in spherical shape of outer diameter 10 mm approximately, wall thickness 0.5 to 0.7 mm. In IFT, the bulb is filled with ^{87}Rb isotope and natural Rb of 99.99% in the 1:1 ratio and each in the quantity of 0.350 mg (aprox). The use of the mixture of natural and ^{87}Rb in equal proportions ensures minimization of the light shift. The cylindrical bulb may also be used. The bulb is filled with Xenon gas at 2.0 ± 0.2 Torr of nearly 99.995% purity for ease of excitation as Xenon has low ionization potential. The only difference for SFT is that, here 700 μgm of pure ^{87}Rb and no natural Rb metal is used. The bulb is excited by 80–100 MHz/3-4 watt RF exciter which is a Clap/Colpitt's oscillator run by a D.C power supply of rating 20 Volt and 0.2 to 0.4 Amp current. The coil, holding the Rb bulb in the tank circuit of the exciter, has 10 turns of enamelled copper wire of 20–22 SWG. **The most important thing to remember is that the coil is wound inside out for concentrating the RF field in the middle, where the Rb bulb is placed**. Otherwise, it is difficult to excite the Rb bulb. The light intensity and mode of the operation of the lamp may be controlled by changing the gain and frequency of the oscillator with the help of a resistor and capacitor respectively. The temperature of the operation is important and should be set at 110°C.

As mentioned and emphasised several times, the Rb lamp is highly critical component, it is necessary to keep provision to controlling lamp exciter gain, frequency and operating temperature of the bulb precisely. To minimize the light shift and to control the light intensity, the light from the lamp is made to pass through a perforated screen. The lamp is self RF heated, but it requires the additional temperature controller, for the stringent stabilization of its intensity and the operation in the proper mode. The temperature plays very important role in the proper functioning of the lamp. When the temperature is around 110°C, the lamp operates in desired ring-mode. The Rb light appears to be emitted from a narrow ring close to the surface of the bulb. The emitted purple coloured Rb light has a very narrow linewidth, and there is no self-reversal. The self-reversal is a phenomenon in which the spectral line shows a depression in the centre of the line profile. This deteriorates the lamp's functioning. When the temperature is increased, the intensity decreases and the colour of the light becomes deep purple or reddish. The spectral lines become broad and highly self-reversed. In view of the above-mentioned behaviour of the lamp at different temperatures, it is inferred that the lamp's temperature should be stabilized at 110°C, using bifilar heater winding for optimum results. The lamp bulb is placed inside a Teflon cylindrical body of inner diameter of 12 mm, length of 20 mm and wall thickness of 2 mm. The one side of the lamp housing is open for the light transmission, while the other side is closed with a stem of length 10 mm projecting out. The Rb bulb notch rests inside a small hole in the stem. This ensures that the bulb notch is relatively cooler and acts as Rb reservoir. The outer surface of the stem has threading, and is screwed to the lamp exciter circuit board. The Teflon casing is placed inside an aluminium (Al) cover, which is heated using a bifilar heater winding. One thermistor sensor for the temperature stabilization is provided on the body of the Rb lamp assembly. On the front side of the Al cover, a glass plate of 1.5 mm thickness is placed to block convection current. This lamp assembly is aligned with the microwave cavity hole, for the transmission of the Rb light through the Rb absorption cell to the photodiode array. The lamp assembly and microwave cavity are jettisoned, using a hollow pipe of Teflon. The lamp exciter circuit is mounted on metal plate, as mentioned in the section on the Rb lamp exciter, and is screwed to the outer most side cover of the Physics package.

Summary

Rb Lamp:

i. For IFT, Spherical bulb outer diameter = 9.3–10 mm, wall thickness = 0.5–0.7 mm, C48-3/Schott 8436/GE-180 glass, Natural Rb + ^{87}Rb, Ratio = 1:1 350 + 350μgm (purity 99.99%), 2 Torr Xenon gas, Operating Temp = 110°C, Exciter power = 3 to 4 watt, frequency = 80–100 MHz.

ii. In the case of SFT, the only difference is that here 700 μgm of high purity ^{87}Rb is used and no natural Rb metal.

10.2.4.2. Rb absorption cell

The Rb absorption cell is also a very critical component of the Physics package like the Rb lamp, mentioned above. The Rb ground state hyperfine atomic transitions take place inside the absorption cell. The line-width or Q of these hyperfine transitions primarily, determines the frequency stability and other characteristics of the Rb atomic clock. In the integrated filter technique (IFT), the filtering of the undesired transitions and also the excitation of the desired hyperfine transitions take place inside the Rb absorption cell itself. In principle, a suitable mixture of ^{87}Rb and ^{85}Rb isotopes in the lamp and the cell should be filled. It is observed from the experience and various experiments, that the desired filtering action may be achieved by using the natural Rb instead of adding ^{85}Rb to ^{87}Rb isotope. A quantity of 1.5 to 2.00 mg of natural Rb is sufficient for the long life of the absorption cell, which is made of Schott 8436/GE-180, or C48-3 glass. Further more, in order to achieve nearly zero light shifts and to minimize the wall relaxation and Doppler broadening, N_2 as a buffer gas at 9.5 Torr pressure is suitable. The Nitrogen gas also provides quenching of the scattered and fluorescence radiation. This property helps in preventing radiation trapping which tends to affect the efficient optical pumping of the Rb atoms. The mechanism behind it is the conversion of the Rb electronic energy into vibration energy of the Nitrogen, and as a result the Rb atoms return to the ground state without the fluorescing, thus avoiding the energy trapping or reabsorbing the fluorescent photons. The absorption cell is kept at a temperature of 75°C for obtaining nearly zero light shifts. It is observed that at 75°C, the filtering temperature coefficient, is negative. Depending on the Rb light intensity, is of the order of a few parts in $10^{-10}\,°C^{-1}$. The Nitrogen buffer gas at 9.5 Torr has positive temperature coefficient of the order of $7\times 10^{-10}\,°C^{-1}$ [9]. These two temperature coefficients of the opposite signs cancel each other to some extent, leaving a residual temperature coefficient of the order of $10^{-10}\,°C^{-1}$ or less.

For SFT, only highly pure ^{87}Rb isotope and the combination of Argon and Nitrogen are used as buffer gases.

Summary

Rb absorption cell:

i. For IFT Cylindrical cell glass 070/7740/C48-3/GE-184, outer diameter = 25 mm, length 27 mm, wall thickness <1mm, Natural Rb 1.5 to 2 mg, Nitrogen buffer gas at 9.5 Torr (purity 99.9995%), Temp = 75°C, temperature stability of fractions of Celsius for frequency stability of 10^{-13} or better.

ii. For SFT, only highly pure ^{87}Rb isotope is used. Buffer gases: Argon and Nitrogen in the ratio (Ar/N2) = 1.538 at 21.7 ± 0.5 Torr and of purity 99.995%.

10.2.5. Resonant microwave cavity

The tuned microwave cavity is an another important component of the Rb atomic clock. If we see the history of development of the Rb atomic clocks, different operating modes of the microwave cavity have been used. Presently, to reduce the size, TE$_{111}$, mode is preferred. A Aluminium (Al alloy 6061) microwave cavity tuned at 6.68 GHz is used for exciting the Rb ground state hyperfine transitions. When developing the microwave cavity for the Rb atomic clock for the first time, it is advisable to begin with a fine treaded tuning plunger. So that the length of the microwave cavity may be precisely determined and fixed. To begin with, its dimensions are L = 40.00 mm, including tuning plunger length of 10 mm, I.D = 27.00 mm and wall thickness 7 mm. The microwave feed is through a loop of 1.00 mm dia on the back wall of the microwave cavity. The photodiode array is also kept inside the cavity. The photodiode array, with ground plane on its backside, acts as the termination of the microwave cavity. The absorption cell completely fills the microwave cavity. The microwave cavity is excited in such a mode of oscillations, that the magnetic field component of RF is in the direction of the optical axis. This is a requirement for the field free clock transitions between $F = 2$, $m_F = 0 \leftrightarrow F = 1$, $m_F = 0$ of ^{87}Rb atoms. These transitions have second order dependence on the applied DC magnetic field. The loaded Q is around 500. The cavity has bifilar heater windings, for maintaining the temperature of the microwave cavity and the absorption cell at 75°C. The required microwave power P is given by the expression

$$P = [h^2 \omega_0 V_c b^2]/2\pi^2 \mu_0 \mu_b^2 \eta Q_c.$$

Here we have, h being the Plank's constant, ω_0 is the RF frequency, V_c is the cavity volume, b the Rabi frequency, μ_0 the permeability, μ_b the Bohr Magnetron, η the filling factor and Q_c the loaded Q of the cavity. The microwave power at the resonant frequency is of the order of 500 Nano watt. In some atomic clocks, to reduce the microwave cavity size considerably, slotted-tube or magnetron-type resonator is used. In this configuration, the cylindrical volume is filled with metal electrodes, which form a combination of capacitance and inductance. The very interesting feature of such a cavity is, that several tuning parameters may be used to obtain the resonance condition very precisely. Physically, the volume can be reduced for the mode of interest, mainly because the balance of electric and magnetic energy associated to the resonance, is achieved through the electrode structure. Typically, such geometries have a reduction of the volume between three and four times compared with the standard cylindrical cavity geometry.

10.2.6. *Magnetic field solenoid*

A constant DC field is applied to the Rb atoms in the absorption cell in order to provide a quantum axis, along the Rb light direction. It also provides some leverage for manipulating the energy of hyperfine levels, so that the hyperfine transition frequency may be matched precisely with the applied microwave frequency. Besides, DC magnetic field may also take care of some uncertainty in the buffer gas pressure. Generally, the magnetic field is of the order of a few hundred milli gauss. The constant DC magnetic field is produced by a solenoid. Its number of turns and the current may be set to produce the required magnetic field. The precision current regulator circuit controls the current in the solenoid.

Summary

Solenoid length, diameter, SWG of winding wire and current are determined by the size of the components of the Physics package. The DC magnetic field range is 0.1 to 0.3 gauss, and the stability of 10 ppm.

10.2.6.1. *Magnetic shield*

The hyperfine transitions are very sensitive to the fluctuations in the external magnetic field. In order to protect the Rb absorption cell and other parts of the Physics package from the external magnetic field fluctuations, three layers of magnetic shield of high magnetic permeability are used. In the case discussed here, the outer most layer has length 200 mm and diameter 130 mm of μ-metal material of very high magnetic attenuation and medium permeability of the order of 2×10^4. The inner two layers are of μ-metal material of medium saturation and high permeability 3×10^4. To achieve the clock frequency fluctuations of less than 10^{-13}, a shielding factor of 10,000 is required. The magnetic shield should be carefully Hydrogen annealed after fabrication and thoroughly demagnetized for the efficient shielding.

Summary

Three concentric cylinders, with shielding factor 10^4 of medium to high permeability (2×10^4 to 3×10^4) μ-metal shield, length of outer most layer and outer diameter are determined by the physical dimensions of the Physics package components.

10.2.7. Base plate

In the space, the heat dissipation is a serious problem. In order to achieve the temperature stability of the various components of the Physics package, it is important to keep thermal balance with the surroundings. This objective is achieved by placing the Physics package on a base plate for the heat dissipation. A base plate temperature controller is provided to reduce the overall temperature sensitivity below the certain level. The base plate also plays important role in reducing the effect of barometric fluctuations on the Physics package. The temperature of the base plate is decided after thorough observations and control of various parameters, affecting the thermal properties of Physics package (please also see Sec. 8.4).

Summary

Material: Al, Size: 150 mm–100 mm, thickness 10–15 mm, heat transfer through radiation and conduction, temperature may be in the range of 25 to 45°C.

10.3. Rb lamp life span determination by calorimeter

The Rb electrode-less lamp is still widely used in space Rb atomic clock and it is very critical component. It determines the life of the Rb clock. Generally, the lamp should have a working life of 10 to 12 years. It is important to know the required mass of Rb metal for total life span at the time of the lamp filling. In initial stages of GPS, the lamp failures resulted in difficult situations. This prompted use of the calorimeter technique for evaluating the rate of consumption of Rb in lamp vs operation time. It is observed that a quantity of 400 μgram Rb in the bulb is necessary for achieving the desired life time of the Rb clock. As the navigation satellite system, the total working life of the on-board the Rb atomic clocks depends largely, on the Rb lamp. In view of the criticality of the life of the Rb lamp, special attention is required on the quantity of Rb metal to be filled. An accurate estimate and measurement of the depletion rate and diffusion to wall of the Rb metal is necessary. This decides the Rb metal fill at the time of preparing the Rb bulb. A method based on the calorimeter is presently used to determine the Rb fill, to ensure its long life spanning 10 to 15 years. The calorimetric measurements are based on the principle of heat flow measurements. The differential calorimetric method is also used for measuring the quantity of the Rb fill, as it is more accurate. In this method, the Rb glass bulb and the identical empty glass bulb, as a reference, are compared, in terms of the heat energy used for melting the Rb metal and the differential heat flow, for the same temperature increase rate. The differential heat flow rate, which is the output of the Differential Calorimeter in calorie/min, is plotted versus time. When the Rb metal starts melting, additional heat is required. The differential heat flow rate at the melting point gives a peak in the plotted curve [252]. On the either side of the melting point, the curve is more or less flat. The integrated area of the peak gives heat required for melting the Rb. There is another parameter, the Rb metal's fusion heat in the unit of calories/gram. This parameter is related to the Rb phase transition. In practical sense, the Rb diffused to the bulb glass does not contribute to Rb phase transition. Therefore, the area in the plotted curve divided by Rb's fusion heat gives information on the quantity of the Rb available in the bulb. If we know the quantity of the Rb initially filled, then we know the rate of the depletion from the differential calorimeter measurements. It is pertinent, at this point, to mention that the depletion rate is also a function of the lamp glass, RF exciter power and temperature of the lamp operation. To minimize the loss of the Rb metal, the bulb is made of Corning 1720/C48-3 or Schott 8436 or GE-180 Alumino-silicate glass. There are availability issues about the Corning and Schott 8436 glass. However, GE-180 is abundantly available as it is used widely in automobile lamps. Therefore, the GE-180 glass is used for Rb bulbs and absorption cells in the present commercial production of Rb atomic clocks. The quantity of the Rb metal in the lamp should be such that it lasts for 10 to 12 years. While filling the Rb bulbs, utmost care is taken about the cleanliness of the bulb shell and purities of Rb metal and buffer gases. In Calorimetric measurements, the Enthalpy or heat energy required for melting Rb is measured in non-destructive test, under known pressure and volume. The calorimeter measurements give good estimate of the lamp working life, and provide information about the quantity of Rb to be filled. It is estimated [252] that in the Rb bulbs, the time dependent consumption rate, follows the equation,

$$C(t) = A + B\sqrt{t}. \tag{10.1}$$

Here, C(t) is the total amount of Rb consumed after the operation for a time t, A is a time independent and fast tiny initial consumption term at the lamp start-up, and B is a diffusive consumption term that lies in the range of 0.1–10 $\mu g/hr^{1/2}$. The values of A and B are to be estimated from the differential calorimetric measurements. The Rb metal consumed versus the lamp operational time plot is fitted to Eq. (10.1) to get the values of the A and B. After several such plots, the average values of A and B are observed to be 0.107 μgram and 0.141 $\mu gram/day^{1/2}$ respectively. Thus from Eq. (10.1), the rate of depletion of the Rb can be known and based on that, the Rb metal is filled for obtaining a life span of 10–12 years.

10.3.1. Helium permeability

The studies show that He permeation through the walls of the Rb lamp and the absorption cell in vacuum or normal atmospheric condition creates aging problems. The negative or positive frequency shift may occur in the Rb atomic clocks. It is observed, that the He permeation problem can be minimized by carefully selecting the glass for the bulb and the absorption cell. The Pyrex 7070 or Schott 8436 or GE-180 type glass shows low Helium permeation. Therefore, either of these glasses, should only be used for the bulbs and the absorption cell in the Rb atomic clocks.

10.4. Electronic package

The Electronic package provides the necessary inputs and controller circuits for the Rb Physics package. Besides, the output from the Physics package for locking the VCXO to the Rb ground state h.f transition frequency is processed by the Electronic package. Its various components are described below.

10.4.1. Heaters with controllers

The temperature control of the Rb bulb and cell to high degree of stability is an important requirement. In the Rb Physics package, the residual temperature coefficient generally is of the order of $10^{-10}\,°C^{-1}$. Therefore, for achieving the Rb atomic clock frequency stability of the order of 10^{-13} or better, the lamp and cell temperatures should have stability of the order of the fraction of a Celsius. To control the temperature of the Rb lamp and the integrated Rb absorption cell, separate heaters are provided. These heaters with proportional current controls have bifilar windings, for eliminating the generation of the magnetic field by the current passing through the heaters. The glass-encapsulated thermistors are placed beneath these heaters for sensing the temperature and providing the necessary input to the temperature controller circuits. In addition to these two heaters, one common heater for maintaining the temperature a few degrees above the ambient temperature is provided. This double oven approach ensures temperature stability of a few milli Celsius. The heaters are DC powered from 18V and 24V power supplies. Besides, a base plate is used and maintained at the constant temperature, to dissipate heat from the Rb Physics package. This helps in thermal stabilization of the space Rb atomic clocks.

10.4.2. Lamp and temperature controller

The Rb lamp is placed in a temperature controlled housing. The lamp temperature controllers housing comprise of thermistor bridges, DC amplifiers and heaters. The heaters are developed from bifilar windings of Kapton foil layer. The temperature is maintained at 110°C

Recipe of Rb Atomic Clock   255

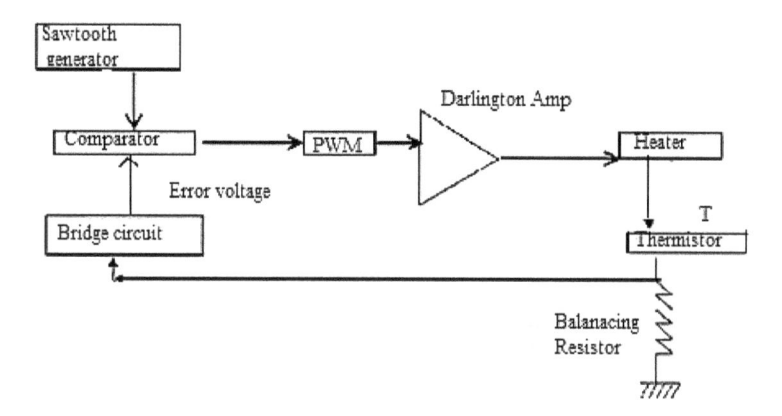

Fig. 10.5 Block diagram of lamp temperature controller.

with the stability of ±0.1°C. The block diagram of the lamp controller circuit is shown in Fig. 10.5.

10.4.3. Servo section

(i) Analog servo section

The function of the servo section is to process the discriminator signal from the Physics package, to produce a frequency control signal. This is done with an analog synchronous detector and an integrator. It generally, includes a 2nd harmonic lock detector and the acquisition sweep circuits. The schematic of the servo loop is shown in Fig. 9.1(b). The photodiode output level is in the range of 0.5 to 10 μA at 137 Hz and synchronous detector output is in the range of 0 to 5 Volt. The levels of the harmonics signal and spurious are −65 and −90 dBc respectively.

(ii) Pre-amplifier

Pre-amplifier section consists of trans-conductance and high gain amplifiers.

- Trans-conductance amplifier

Trans-conductance is a low noise bipolar op-amp. A resistor is used in the feedback, providing a trans-conductance bandwidth in the range of a few KHz. The input current noise (in the range of pA/$\sqrt{\text{Hz}}$) of the op-amp and the Johnson noise current (in the range of pA/$\sqrt{\text{Hz}}$) of the feedback resistor are not critical. The shot noise current times the trans-conductance gain is in the range of 100–1000 nV/$\sqrt{\text{Hz}}$.

- *High gain amplifier*

This amplifier is designed with low noise bipolar input op-amp. This amplifier raises the level of the output voltage of trans-conductance amplifier. The input level may be in the range of 50 to 100 μV. The gain is decided according to the desired output level. The gain band-width is less than 5 MHz.

- *Notch filter*

A notch filter is used for rejecting the second harmonics of the modulating 137 Hz signal from the detected signal. The second harmonic rejection level may be > 30 dBc with the filter band-width of 10 to 15 Hz.

(a) *Digital servo section*

The analog servo may be replaced by a digital circuit, which offers a simpler implementation and eliminates errors. A first step in that direction is to use an Integrate and Dump circuit at the front end. Digitize the resulting discriminator information and process it numerically, finally convert the digital output back to an analog VCXO control voltage. It is the best applied for relatively slow modulation rates, where there is no large 2nd harmonic content and the transients can be suppressed by the reset interval. The Integrate and Dump servo block diagram is given Fig. 10.6.

10.4.4. *Lock-in amplifier*

The lock-in amplification is to separate a small, narrow-band signal from the interfering noise. The lock-in amplifier acts as combination of a phase sensitive detector and a narrow-band filter. The very small signals can be detected in the presence of the noise which may be

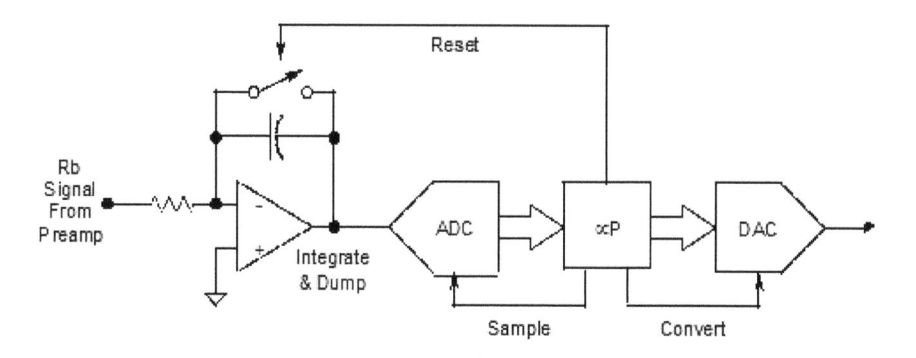

Fig. 10.6 Integrate and Dump servo section.

several times of the signal level. The lock-in amplifier is, basically, a synchronous demodulator followed by a low-pass filter.

10.4.5. *Synchronous demodulation*

The output of balance demodulator is the base band modulation signal. The higher order carrier components, that can be removed with a low-pass filter, are also present. The dynamic range is greater than 90 dB. The input and reference frequencies are same 137 Hz with some phase difference, the output of the synchronous demodulator is DC and riding on it is second harmonic 274 Hz.

10.4.6. *137 Hz oscillator, integrator, lock-detector and sweep circuit*

The modulating frequency, 137 Hz is generated by dividing 8.768 KHz frequency by factor of 64 using CD 4060 CMOS oscillator divider IC. There is no particular requirement on the stability of this frequency generation, as any frequency drift or error is nullified at the homodyne type of demodulation, i.e., any error occurring in the both the inputs i.e., one coming from the photodiode and other from the oscillator, cancel out each other.

A low pass filter reduces noise at the output, which contains DC and second harmonic 274 Hz. The bandwidth of the integrator decides the noise suppression. A bandwidth of 400 Hz and gain of 100 dB are required for processing the clock signal.

A second harmonics from the trans-conductance amplifier detects the locking to the resonance signal, from the Physics package, of the 10 MHz OCXO/VCXO.

A high frequency switch is used to control the sweep range. Its output is either low or high when no lock is present. The output of switch circuit is routed to the integrator input.

10.4.7. RF section

The RF section generates the required 6.834 GHz frequency from the 10 MHz ovenized VCXO. There are different topologies available for this function, which are discussed in the following sub-sections.

10.4.7.1. *Analog frequency synthesis*

For all the LOs, there are two possible topologies to generate the required output frequency from 10.23 MHz reference signal.

(*i*) *Frequency multiplier approach*

In this approach frequency multiplier stages are cascaded to achieve the final output frequency. After each multiplier stage, filter and amplifier are used to clean up unwanted harmonics, generated by the multiplier. Then the desired signal is amplified to a sufficient level, which can drive the subsequent multiplier stage, Fig. 10.7.

(*a*) *All digital servo section*

A further step toward RFS servo digitalization is to directly sample the discriminator signal, completely process it numerically, and convert to an analog VCXO control voltage. An all-digital servo

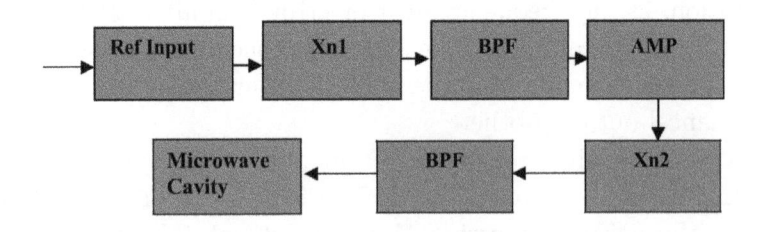

Fig. 10.7 RF section block diagram using frequency multiplier topology.

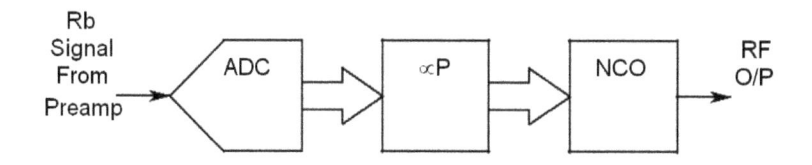

Fig. 10.8 All digital servo system.

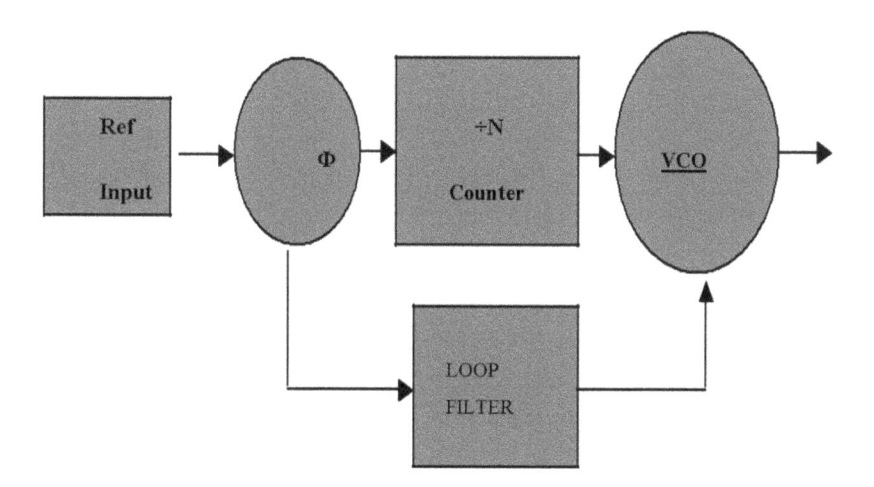

Fig. 10.9 RF section block diagram using PLL topology.

applies the digital frequency control information directly to a numerically controlled oscillator (NCO). The block diagram is shown in Fig. 10.8.

(ii) *Phase locked loop approach*

In this configuration, the respective LO frequencies are directly generated by a VCO and phase locked to the reference signal, after frequency division by a suitable factor. PLL approach has the advantage of less RF hardware as shown in Fig. 10.9.

(iii) *Other option*

Here we consider the analog RF chains that generate the Rb hyperfine frequency 6.834682611 GHz by multiplying a 10 MHz crystal oscillator to obtain 6840 MHz, and subtracting an offset of 5.3125 MHz in a mixer. This approach has a large buffer gas offset (~4.9 kHz), and had poor microwave spectral purity. Better microwave spectral purity can be obtained by straight multiplication of an exact sub-multiple of the Rb frequency, usually from a phase-locked crystal oscillator. The multiplier chain is selected in the first phase of designing.

(a) *Resonance microwave frequency 6.8346875 GHz generation chain*

The RF section is for generating resonant microwave 6.8346875 GHz signal from the 10 MHz coming from the VCXO. The RF generation section consists of an active multiplier, amplifier, filter and SRD diode multiplier, as given in Fig. 10.10. The specifications are given in Table 5.5.

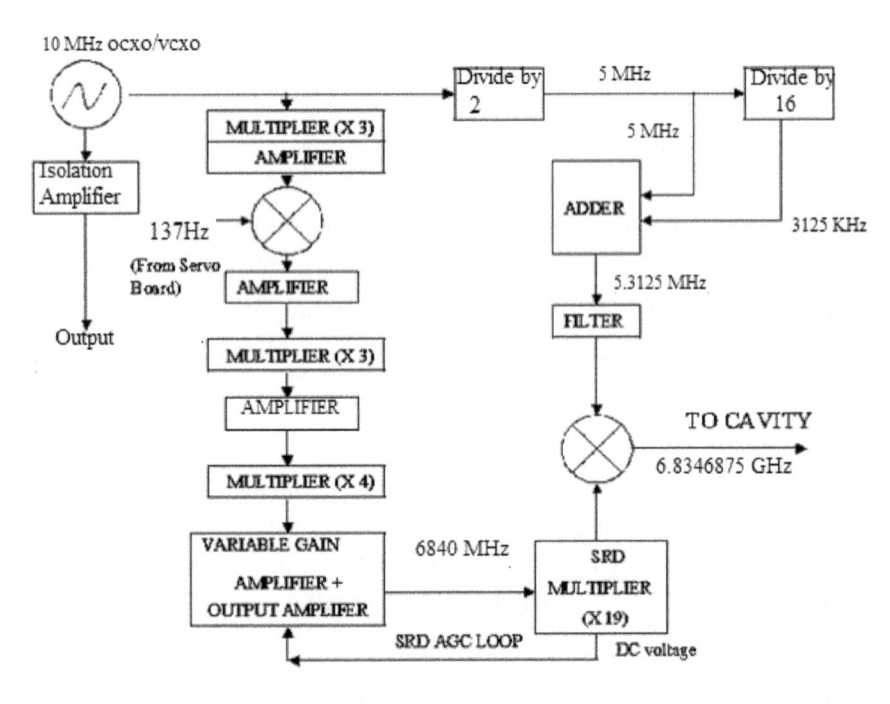

Fig. 10.10 Schematic of RF generation section.

(b) *Spectral purity and offset*

Some Experiments show that, for small microwave purity, larger offset is acceptable that allows simpler frequency synthesis schemes. The expected 4.6 kHz buffer gas offset is handled in some cases by measuring the cell resonance frequency as it is filled with buffer gas.

- *10 MHz VCXO*

10 MHz VCXO with power output of 3 dBc in 50 ohm, is used to generate 6.8346875 GHz frequency for exciting the resonance cavity. The corrected output of 10 MHz is also the Rubidium clock output. It has the tuneability of ±1.0 Hz and the aging rate is 30 ppb/year with the short-term stability of 1×10^{-12} for a sample time 1s. The harmonics and spurious levels are supposed to be below −30 and 60 dBc respectively.

- *10 MHz × 3 multiplier*

This multiplier generates the 3rd harmonic of its input frequency and it is designed using RF BJT (BFY-193). The input power is 0 dBm, conversion gain 5 dB and harmonics less than 20 dBc. This circuit is designed on 1.6 mm Glass epoxy PCB and its simulated response in shown in Fig. 10.11.

- *10 MHz amplifier*

This amplifier is used to increase the output power of the RF signal to drive varicap diode. This amplifier has transistor 2N2222A as an active component and is implemented on 1.6 mm Glass epoxy PCB. The input and output powers are 0 and 10 dBm respectively and harmonics are maintained below 20 dBc.

- *137 Hz phase modulator*

The varicap diode along with primary of transformer, work as a tuned tank circuit. The tank reactance phase modulates 10 MHz frequency, which is coupled to the next stage of the amplifier using secondary winding of the transformer. The modulation index decides the depth

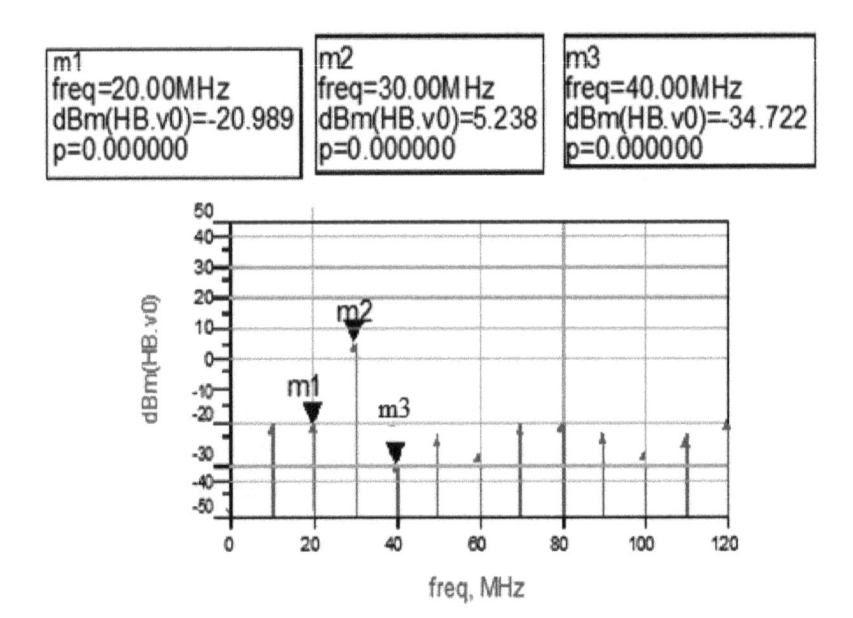

262   Rubidium Atomic Clock: The Workhorse of Satellite Navigation

```
m1                      m2                     m3
freq=20.00MHz           freq=30.00MHz          freq=40.00MHz
dBm(HB.v0)=-20.989      dBm(HB.v0)=5.238       dBm(HB.v0)=-34.722
p=0.000000              p=0.000000             p=0.000000
```

Fig. 10.11   Simulated response of 10 MHz × 3 multiplier.

of modulation. That is dependent on the amplitude of modulating frequency i.e., if we have requirement of ±5 KHz depth of sweep in the modulation than the correspondingly modulation index at 10 MHz is

$$\beta = \Delta f/f_m = (10\,\text{KHz}/600{*}127) \sim 0.1.$$

- *30 MHz amplifier*

This amplifier is used for raising the output power of the RF signal to drive the next multiplier stage circuit. This amplifier is designed with the BJT, **BFY-193** and implemented on 1.6 mm Glass epoxy PCB. The input and output powers are kept 0 dBm with harmonics suppression to 20dBc.

- *30 MHz × 3 multiplier*

This multiplier generates 90 MHz, the 3rd harmonic of its input 30 MHz frequency and is designed using RF BJT (BFY-193). The conversion gain is 5 dB with harmonics below 20 dBc. This circuit is

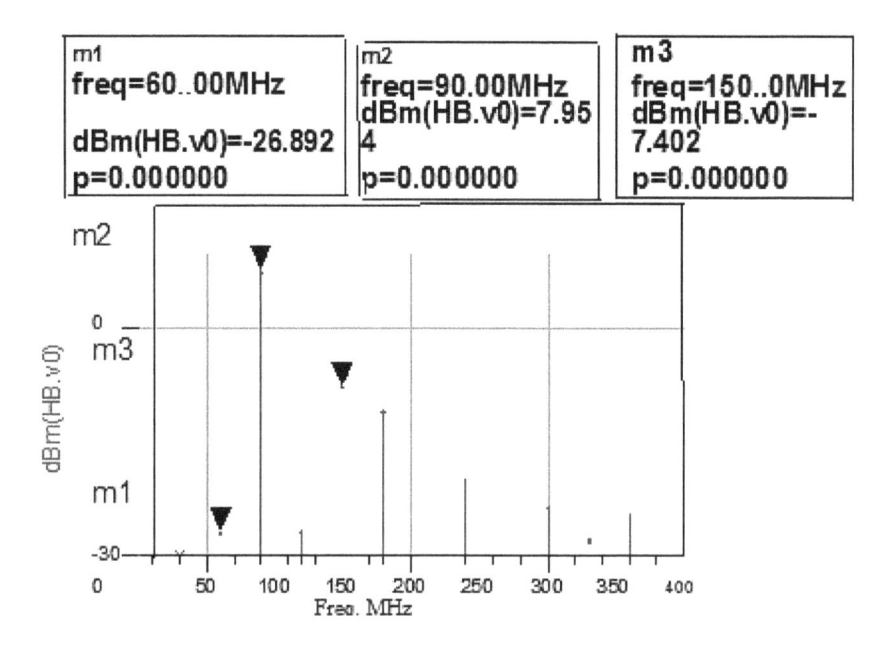

m1	m2	m3
freq=60..00MHz	freq=90.00MHz	freq=150..0MHz
dBm(HB.v0)=-26.892	dBm(HB.v0)=7.954	dBm(HB.v0)=-7.402
p=0.000000	p=0.000000	p=0.000000

Fig. 10.12   Simulated response of 30 MHz × 3 multiplier.

designed on 1.6 mm Glass epoxy PCB and the simulated response is shown in the Fig. 10.12.

- *90 MHz crystal filter and amplifier*

This filter is used to reject unwanted harmonics below 75 dB at 90 MHz ±10 MHz, generated by the ×3 active multiplier. The insertion loss is about 5 dB. The crystal filter package is mounted on 1.6 mm Glass epoxy PCB. The filter is followed by an amplifier. It raises the level of 90 MHz signal available at the O/P of the crystal filter, to drive the next multiplier stage with the gain of 15 dB. This amplifier is designed with the RF-BJT, BFY-193.

- *90 MHz × 4 multiplier*

This multiplier is used to generate the 4th harmonic of the input 90 MHz frequency. It is designed using BJT (BFY-193) as an active device. The specifications: Input and ouput frequencies 90MHz and

360 MHz respectively, input power 0dBm with conversion gain 1 dB and harmonics less than 20 dBc. This circuit is fabricated on 1.6 mm Glass epoxy PCB.

- *Variable gain amplifier*

The variable gain amplifier is used to maintain a constant RF level. The step recovery diode produces a DC bias voltage. This DC bias voltage is monitored and is compared with the predefined voltage set by voltage divider and a potentiometer. The bias of the variable gain amplifier changes according to the error signal, generated from the SRD.

- *360 MHz amplifier*

This amplifier is used to increase the output power level of the RF signal to drive the SRD diode. The specifications: output power 20 dB and in/out return loss >20 dB.

- *360 × 19 SRD multiplier*

The step recovery diode is a varactor diode designed for the frequency multiplication to 6840 MHz. The resonator cavity is tuned to select the 19th harmonics of 360 MHz, which corresponds to the resonant frequency of the Rb. The input and output powers are 20 and 0 dBm respectively. With no specific SRD model, we do the simulation by putting the specifications of HP 0320 SRD device parameters in the ideal SRD model. The results are shown in Fig. 10.13. The parameters of the selected SRD are CVR : 2 pF, Rs : 0.4 ohms and τ : 44 ns.

- *10 MHz ÷ 2 frequency divider*

A divide by 2 circuit is used to get 5 MHz from the 10 MHz VCXO signal. This circuit is realized using 74HC74. The input and output levels are 2 and 4 p-p volts respectively.

- *5 MHz ÷ 16 frequency divider*

To meet the requirement of Telecom, the frequency 3.125 MHz is generated by dividing 5 MHz signal by a factor of sixteen. This oscillator

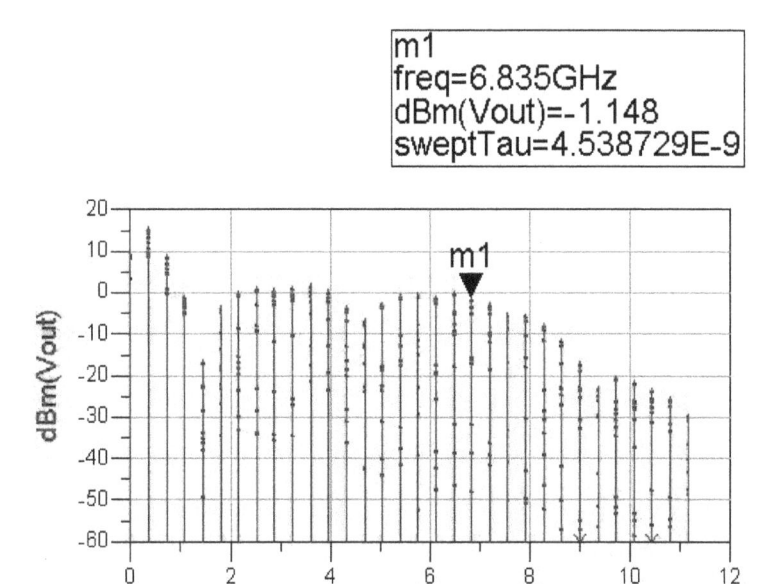

```
m1
freq=6.835GHz
dBm(Vout)=-1.148
sweptTau=4.538729E-9
```

Fig. 10.13 Simulated response of SRD multiplier.

circuit has been realized using 74HC74. The input and output levels are 4 p-p volt each.

- *Mixer for 5 MHz with 3.125 KHz*

The frequency 6840 MHz from SRD is mixed with synthesized 5.3125 MHz to get the exact resonance frequency. The mixer circuit is used for adding the 5 MHz with the 312.5 KHz. This circuit is realized using 74HC86 XOR gate. The input and output levels are more than 4 volt p-p each, with phase noise at 1KHz and 100 KHz being kept below −110 and −120 dBc respectively.

- *5.3125 MHz crystal filter*

This filter is used to reject unwanted harmonics generated by the frequency adder. The filter insertion loss is less than 5 dB at the frequency, 5.3125 MHz. The spurious rejection at ±10.23 MHz about the signal frequency is more than 75 dB. The crystal filter may be mounted on 1.6 mm Glass epoxy PCB.

- Balance mixer

This mixer is use to mix the two input frequencies 6840 and 5.3125 MHz coming from the SRD and then to select the lower sideband to generate the resonant frequency 6834.6875 MHz. As the cavity requires the microwave power of 500 nano watt = −33 dBm, to regulate, a −30 dB attenuation is provided by using thermopad/attenuators after the frequency adder.

(*iv*) *PLL approach*

Better microwave spectral purity can be obtained by straight multiplication of an exact sub-multiples of the Rb frequency, usually from a phase-locked crystal oscillator, Fig. 10.14.

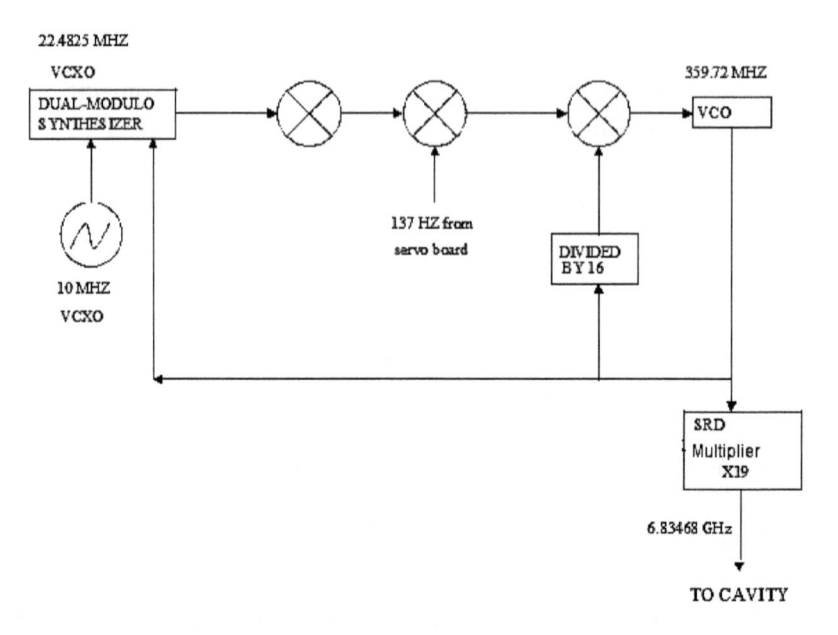

Fig. 10.14 RF section using PLL approach.

(v) *Digital frequency synthesis*

The Direct Digital Synthesizer (DDS) greatly simplifies development of RF chains, making the absorption cell buffer gas offset a free variable. Thus removing its tight fill tolerance, and offering an ideal way to apply servo modulation. As long as the DDS signal is inserted additively, 32-bit resolution is adequate for the fine-tuning.

10.4.7.2. *Electronic power conditioner (ECP)*

ECP is required to provide adequate power to various sub-systems in the Rb atomic clock for its proper functioning and electrical interface to the navigation satellite. It contains DC to DC converters. The input voltage is 27 to 45 volt. The power supply is regulated to eliminate over and under voltage with the inbuilt protection mechanism.

10.4.7.3. *Miniaturization of Rb atomic clock*

Rb atomic clock may be miniaturized to a considerably small size. However, the tuned microwave cavity decides the overall size of the Rb atomic clock. It is pertinent to mention that the mode of microwave cavity should be such that its RF magnetic field and the optical pumping radiation are in the same direction. Initially, microwave cavity in TE_{011} mode was used as it had good Q and losses were less. The constraint on cavity was its large size. To reduce the size, the microwave cavity operating in higher mode TE_{111} was introduced. In most of the commercial Rb clocks, this mode is preferred. Therefore, in this book this mode has been discussed in detail. However, R&D work on microwave cavity for the Rb atomic clock, has resulted in the development of loop-gap microwave cavity, shown in the Fig. 10.15. In the conventional Rb clocks, the size of the Rb cell is very small to the volume of the microwave cavity. The microwave cavity determines the overall size of Rb clock. A very efficient way to reduce the dimensions is to use loop-gap or slotted tube or magnetron microwave cavity [253]. The basic concept to reducing the size, is that for a given frequency, it is not the size but the balance between magnetic and electrical fields. It is achieved through electrodes resulting in size reduction to $1/4^{th}$ or $1/5^{th}$ of the conventional cavity.

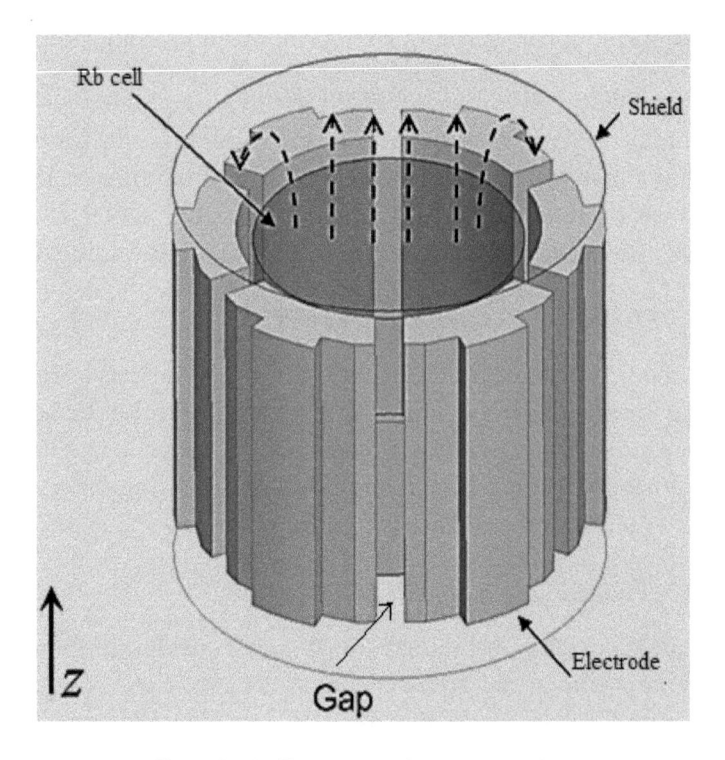

Fig. 10.15 Loop-gap microwave cavity.

The good aspect of the loop-gap cavity is that the homogeneity of electromagnetic field is not degraded. The metal electrodes provide the frequency determining inductance and capacitance. The loop-gap cavity resonance condition can be fulfilled more precisely. The loop-gap microwave structure in Fig. 10.15 is based on six electrodes. Two empty cylindrical extensions are situated on both sides of the loop-gap region. The dashed arrows represent the RF magnetic field lines. The outer circle is the shield. This type of microwave cavity may reduce the size of the Rb atomic clock. There is another highly miniaturized model of Rb which is of chip size. It is based on the principle of coherent population trapping (CPT) [254]. The commercial manufacturing of chip scale Rb atomic clocks has been undertaken by the manufacturers of the time and frequency test and measurement instruments.

Conclusion

The complete description of the role of Rb atomic clocks in the modern day satellite navigation systems has been the objective of the book. The given details on Rb atomic clock are quite exhaustive in technicalities. The authors are of the opinion that the book will find acceptability amongst readership right from the experts to those new to the subject. The presentation in the book has taken care of including all or most of the relevant information on the subject. The main emphasis is on the practical aspects of Rb atomic clocks for the readership who wish to know, how to develop these Rb clocks as the practical devices. The R & D work on the Rb atomic clocks has been extensively done for more than six decades. Even then, intense research interest in Rb atomic clocks may be seen at the present time. Therefore, the publications on Rb clocks are available in large numbers. However, the critical information on their development was a missing link. The authors have sincerely put efforts to bridge that missing link. The procedure of developing Rb isotopic bulbs, filters and absorption cells is very intricate and rarely described. In Chapter 10 of the book, the Rb bulbs and cells development has been described with emphasis on their practical realization. If this book can generate amongst the readers any interest in the atomic clocks, the authors will have the satisfaction that their efforts are rewarded. Those who want to go deeper in the subject can refer to the cited publications in the bibliography.

Bibliography

1. J. Jespersen and J. Fitz-Randolph. From Sundials to Atomic Clocks: Understanding Time and Frequency. U.S. Department of Commerce, Washington, D.C., 1999. (National Institute of Standards and Technology, Monograph 155, (1999).
2. C. Hackman and D. B. Sullivan. Resource letter: TFM-1: Time and frequency measurement, *American Journal of Physics*, **63**, no. 4, pp. 306–317, (1995).
3. I. I. Rabi, J. R. Zacharias, S. Millman, and P. Kusch. "A new method of measuring nuclear magnetic moment", *Physical Review*, **53**, no. 4, p. 318, (1938). doi:10.1103/PhysRev.53.318.
4. The atomic clock: An atomic standard of frequency and time. *National Bureau of Standards Technical News Bulletin*, **33**, no. 2, pp. 17–24, (1949). Available from: http://tf.boulder.nist.gov/general/publications.htm.
5. M. A. Lombardi, T. P. Heavner, and S. R. Je_erts. NIST primary frequency standards and the realization of the SI second. *Measure: The Journal of Measurement Science*, **2**, no. 4. pp. 74–89, (2007). Available from: http://www.nist.gov/publication-portal.cfm.
6. T. Quinn, "Fifty years of atomic time-keeping: 1955 to 2005", *Metrologia*, **42**, no. 3, (2005).
7. W. Markowitz, R. Glenn Hall, L. Essen, and J. V. L. Parry, "Frequency of cesium in terms of ephemeris time". *Physical R Review Letters*, **1**, no. 3, pp. 105–107, (1958).
8. Resolution 1 of the 13th Conference Generale des Poids et Mesures (CGPM), (1967).
9. J. Vanier and C. Audoin. Rb frequency standards, Chapter 7, The Quantum Physics of Atomic Frequency Standards. A. Hilger, Bristol, (1989).
10. S. Lecomte, M. Haldimann, R. Ruffieux, P. Berthoud, and P. Thomann. "Performance demonstration of a compact, single optical

frequency cesium beam clock for space applications". In *IEEE Frequency Control Symp. and 21st European Frequency and Time Forum*, pp. 1127–1131, Geneva, Switzerland, (2007).
11. R. Wynands and S. Weyers. "Atomic fountain clocks". *Metrologia*, **42**, no. 3, June (2005).
12. Savita Singh, Bikash Ghosal and G. M. Saxena, Nov.2010, arXiv:1011.2355v1 [physics.gen-ph]
13. C. AffoldeRbach, F. Droz, and G. Mileti. "Experimental demonstration of a compact and high-performance laser-pumped Rb gas cell atomic frequency standard", *IEEE Trans. Instrum. Meas.*, **55**(429), (2006).
14. J. Vanier and C. Mandache. "The passive optically pumped Rb frequency standard: The laser approach". *Applied Physics B: Lasers and Optics*, 87:565–593, (2007).
15. S. A. Diddams, Th. Udem, J. C. Bergquist, E. A. Curtis, R. E. Drullinger, L. Hollberg, W. M. Itano, W. D. Lee, C. W. Oates, K. R. Vogel, and D. J. Wineland. "An optical clock based on a single trapped 199Hg+ ion". *Science*, **293**(5531):825–828, (2001). doi:10.1126/science.1061171.
16. M. Takamoto, F.-L. Hong, R. Higashi, and H. Katori. "An optical lattice clock", *Nature*, **435**(7040):321–324, (2005). doi:10.1038/nature03541.
17. R. H. Dicke, "The effect of collisions upon the Doppler width of spectral lines," *Phys.Rev.* **89**, pp. 472–473; (January 15, 1953).
18. H. Fruhauf and W. Weidemann, "Development of a Sub-miniature Rb Oscillator for SEEKTALK Application", *Proceedings of the Twelfth Annual Precise Time and Time Interval (PTTI) Application and Planning Meeting*, (Dec. 1980).
19. J. Vanier, L. G. Bernier, A. Brisson, M. Tetu, and J. Y. Savard, "Possible avenues of improvement of the short and long term stability of optically pumped passive Rb frequency standards", *Proc. 34th Ann. Freq. Control Symp*, p. 376, (1980).
20. A. Risley, S. Jarvis Jr., and J.Vanier, "The dependence of frequency upon microwave power of wall-coated and buffer-gas-filled gas cell 87Rb frequency standards", *J. Appl. Phys.*, **51**, 4571, (1980).
21. W. E. Bell, A. L. Bloom, and J. Lynch, "Alkali metal vapour spectral lamps". *Review of Scientific Instruments*, **32**(6): pp. 688–692, (1961).
22. R. W. Shaw, "Spontaneously Generated Ion Acoustic Waves in Weakly Ionized Plasma," Master Thesis, Cornell University, Ithaca, (1964).

23. J. Vanier and C. Mandache "The passive optically pumped Rb frequency standard: The laser approach", *Appl. Phys. B*, **87**, pp. 565–593, (2007).
24. F. X. Esnault, et al. "HORACE: A compact cold atom clock for Galileo", *J. Adv. Space Res*, **47**(5), pp. 854–858, (2011), doi:10.1016/j.asr.2010.12.012.
25. P. Rochat, et al., "The onboard Galileo Rb and passive maser, status and performance", *Proc. IEEE Frequency Control Symp.*, Vancouver, BC, USA, pp. 26–32, (2005).
26. S. Schiller, et al., "The space optical clock project". *Proc. Int. Conf. on Spacet Optics*, Rhodes, Greece, (2010).
27. J. Camparo. "The Rb atomic clock and basic research". *Physics Today*, pp. 33–39, Nov. (2007).
28. L. A. Mallette, P. Rochat, and J. White, "An introduction to satellite based atomic frequency standards", *IEEE Aerospace Conference*, pp. 1–9, Big Sky, MT, May (2008).
29. L. A. Mallette. "Atomic and quartz clock hardware for communication and navigation satellites". In *39th Annual Precise Time and Time Interval (PTTI) Meeting*, (2007).
30. P. Waller, S. Gonzalez, S. Binda, I. Sesia, I. Hidalgo, G. Tobias, and P. Tavella. "The in-orbit performances of GIOVE clocks". *IEEE Trans. Ultrason. Ferroelctr. Freq. Control.* **57**:738, (2010).
31. N. F. Ramsey, "Applications of atomic clocks," in Laser Physics at the Limits, H. Figger, D. Meschede, and C. Zimmermann, eds. (Springer, Berlin), pp. 3–8, (2002).
32. J. D. Prestage, R. L. Tjoelker, and L. Maleki. "Atomic clocks and variations of the fine structure constant". *Phys. Rev. Lett.*, **74**, no. 18, pp. 3511–3514, (1995).
33. J. Levine, "Time synchronization using the internet", *IEEE Trans. UFFC*, **45**, 450, (1998).
34. J. Levine, M. Weiss, D. D. Davis, D. W. Allan, and D. B. Sullivan, "The NIST automated computer time service", *J. Res. NIST*, **94**, 311, (1989).
35. D. Jackson and R. J. Douglas, "A telephone-based time dissemination system", *Proc,18th Ann.PTTI meeting*, available from USNO time service, 541, (1989).
36. M. Rohden, A. Sorge, M. Timme, and D. Witthaut. "Self-Organized Synchronization in Decentralized Power Grids". *Phys. Rev. Lett.*, **109** (064101): 1–5, (2012).
37. M. A. Lombardi, T. P. Heavner, and S. R. Je_erts. NIST primary frequency standards and the realization of the SI second. Measure: *The Journal of Measurement Science*, **2**, no. 4, pp. 74–89, (2007).

38. W. M. Itano and N. F. Ramsey. "Accurate measurement of time". *Scientific American*, **269**:56–65, (1993). 1, 140.
39. D. W. Allan, N. Ashby, and C. C. Hodge. "The Science of Timekeeping". Technical report, Hewlett-Packard, (1997).
40. J. F. Pascual-Sánchez. "Introducing relativity in global navigation satellite system". *Annalen der Physik*, **16**:258–273, (2007).
41. M. Bird, M. Allison, S. Asmar, D. Atkinson, I. Avruch, R. Dutta-Roy, Y. Dzierma, P. Edenhofer, W. Folkner, D. Johnston, D. Plettemeier, S. Pogrebenko, R. Preston, and G.Tyler. "The vertical profile of winds on Titan". *Nature*, **438**:800–802, (2005).
42. M. Antonello *et al.* "Measurement of the neutrino velocity with the ICARUS detector at the CNGS beam". *Phys. Lett. B*, **713**:17–22, (2012).
43. A. Bloom, *Sci.Am.* **203**, 72 (1960); W. Happer, W. van Wijngaarden, Hyperfine Interact. **38**, 435, (1987).
44. P. Bender, E. Beaty, A. Chi, *Phys. Rev. Lett.*, **1**, pp. 311, (1958).
45. M. Arditi and T.R. Carver, "The principles of the double resonance method applied to gas cell frequency standards". *Proceedings of the IEEE*, 1963, **51**, no. 1, pp. 190–202, (1963).
46. J. M. Anderson, "Electrodeless Fluorescent Lamps Excited by Solenoidal Electric Fields", *National Technical Conference of the Illuminating Engineering Society*, Phoenix, Arizona, September 9–12, (1968).
47. R. B. Piejak, V. A. Godyak, and B. M. Alexandrovich, "A simple analysis of an inductive RF discharge", *Plasma Sources Sci. Technol*, **1**, pp. 179–186, (1992).
48. L. R. Nerone and A. H. Qureshi, "Mathematical modelling and optimization of the electrodeless, low-pressure, discharge system", 24th *Annual IEEE Power Electronics Specialist Conference*, University of Washington, Seattle, pp. 509–514, (1993).
49. F. Paschen, "Über die zum funkenübergang in luft, wasserstoff und kohlensäure bei verschiedenen drücken erforderliche potentialdifferenz", *Weid. Annalen der Physik*, **37**, pp. 69–75, (1889).
50. Pozar David M, Microwave Engineering, 3rd Ed, John Wiley & Sons Inc., pp. 170–174, (2005).
51. E. Nasser, "Fundamentals of gaseous ionization and plasma electronics", *Wiley-Interscience*, (1971).
52. Lighting Research Center, "Electrodeless Lamps, The next generation," *Lighting Futures*, **1**, no. 1, May/June (1995).
53. D. O .Wharmby, "Electrode-less lamps for lighting: A review", *IEE Proceedings-A*, **140**, no. 6, pp. 465–473, (1993).

54. J. W. Shaffer and V. A. Godyak, "The development of low frequency, high output electrodeless fluorescent lamps," *Journal of the IES*, **28**, no. 1, pp. 142, (1999).
55. Siemens. Application Note 023, Edition A01. October 17, (2006).
56. A. Goel and H. Hashemi, "Frequency switching in dual-resonance oscillators," *IEEE Journal of Solid-State Circuits*, **42**, no. 3, pp. 571–582, (2007).
57. Agilent Technologies, Advanced Design System, Agilent EEs of EDA, http://eesof.tm.agilent.com/docs/adsdoc2003A/manuals.htm, (2003).
58. Lighting Research Center, "Electrode less lamps, The next generation", *Lighting Futures*, **1**, no. 1, June (1995).
59. D. O. Wharmby, Review of electrodeless discharges for lighting', *Proceedings of 5th international symposium on the science and technology of light sources*, University of Sheffield, (1989).
60. J. W. Shaffer and V. A. Godyak, "The development of low frequency, high output electrodeless fluorescent lamps," *Journal of the IES*, **28**, no. 1, p. 142, (1999).
61. R. B. Piejak, V. A. Godyak, and B. M. Alexandrovich, "A simple analysis of an inductive RF discharge," *Plasma Sources Sci. Technol.* **1**, pp. 179–186, (1992).
62. J. N. Lester and B. A. Alexandrovich, "Ballasting electrodeless fluorescent lamps", *Journal of the IES*, **29**, no. 2, pp. 89–99, Summer (2000).
63. Osam Sylvania INC., "ICETRON, Fixture Design Guide 100 Endicott St.," Danvers, MA, 01923.
64. N. P. Ivanov, "The atomic-absorption analysis, Method of chemical reactions and preparations", **10**, pp. 21–23, (1965).
65. B. V. Levov, "The Atomic–Absorption Spectral Analysis", Nauka, Moscow, (1966).
66. A. J. Skudra, "Research and development of high-frequency electrodeless lamps of helium and mercury", book, University of Riga, (1992).
67. S. Tolanski, High Resolution Spectroscopy, Mmethuen, London, (1955).
68. W. F Meggers and F. O Westfall, "Lamps and wave-lengths of mercury198", *J. Res. NBS*, **44**, pp. 447–455, (1950).
69. N. A. Kaptsov, "The Electrical Phenomena in Gasses and in Vacuum", Nauka, Moscow, (1953).
70. C. Corliss, W. Bozman, and F. Westfall, "Electrodeless metal–halide lamps", *J. Opt. Soc. Amer.*, **43**, pp. 398–400, (1953).

71. J. Le Bel, "An electrical lamp", Patent no. US 2118452, May 24, (1938).
72. J. C. Camparo and C. M. Klimcak, "Discharge lamp stabilization system", Patent no. US 7,221,231 B2, May 22, (2007).
73. J. Vanier, R. Kunski, A. Brisson, and P. Paulin, "Progress and prospects in Rb frequency standards", *Journal De Physics*, **42**, (1981).
74. J. C. Camparo and R. MacKay, "Spectral mode changes in an alkali RF discharge," *J. Appl. Phy.*, **101**, 053303, (2007).
75. R. W. Shaw, "Spontaneously generated ion acoustic waves in weakly ionized plasma", Master book, Cornell University, (1964).
76. T. Colbert and J. Huennekens, "Radiation trapping under conditions of low to moderate line center optical depth," *Phys. Rev. A*, **41**, pp. 6145–6153, (1990).
77. Y. Q. Wang, J. S. Fu, T. Q. Dong, *et al.* Principle of Quantum Frequency Standard (in Chinese). *Beijing: Science Press*, **391**, (1986).
78. S. C. Brown, "Basic data of plasma physics: The fundamental data on electrical discharges in gases", *American Institute of Physics*, (1967).
79. V. V. Gershun, V. I. Khutorshchikov, and N. N Jacobson, "Shift of the Rb line 794.8nm by extraneous gases", *Opt. Spectros.*, **31**, pp. 866–869. (1971).
80. V. Bell, A. Bloom, and J. Linch, "Spectral lamps of alkaline metals filled by vapour", *Rev. Sci. Instrum.*, **32**, pp. 558–583, (1961).
81. Q. Q. Sun, X. Y. Miao, R. W. Sheng, *et al.* "The near-infrared spectra and distribution of excited states of electrodeless discharge Rb vapour lamps", *Chin. Phys. B.*, **21**, 033201, (2012).
82. V. B. Gerard, "Laboratory alkali metal vapour lamps for optical pumping experiments," *J. Sci. Instrum.*, **39**, pp. 217–218, (1962).
83. J. C. Camparo and R. Mackay, "Spectral mode changes in an alkali RF discharge," *J. Appl. Phys.*, **101**, 053303, (2007).
84. W. E. Bell, A. L. Bloom, and J. Lynch, "Alkali metal vapour spectral lamps," *Rev. Sci. Instrum.*, **32**, pp. 688–692, (1961).
85. R. G. Brewer, "High intensity low noise Rb light source," *Rev. Sci. Instrum.*, **32**, pp. 1356–1358, (1961).
86. D. J. Wineland, J. C. Bergquist, J. J. Bollinger, W. M. 1-0, D. J. Heinzen, S. L. Gilbert, C. H. Manney, and M. G. Raizen, "Progress at NIST toward absolute frequency standards using stored ions", *IEEE Trans. Ultrasonics, Ferroelectrics, and Frequency Control*, **37**, No. 6, pp. 515–523, Nov. (1990).
87. D. Prestage, G. J. Dick, and L. Maleki, "JPL Trapped ion frequency standard development", *IEEE 41st Annual Frequency Control Symposium*, pp. 20–24, (1987).

88. M. G. Raizen, J. M. Gilligan, J. C. Bergquist, W. M. Itano, and D. J. Wineland, "Ionic crystals in a linear paul trap", *Physical Review*, **45**, no. 9, pp. 6493–6501, May 1, (1992).
89. A. de Marchi, "The optically pumped caesium fountain: 10^{-15} frequency accuracy", *Metrologia*, **18**, pp. 103–116, (1982).
90. A. Clairon, C. Salomon, S. Guellati, and W. D. Phillips, "Ramsey resonance in a zacharias fountain", *Europhysics Letters*, **16**(2), pp. 165–170, Sept. (1991).
91. S. Chu, "Laser manipulation of atoms and particals", *Science*, **253**, pp. 861–866, (1991).
92. J. L. Hall, M. Zhu, and P. Buch, "Prospects for using laser-prepared atomic fountains for optical frequency standards applications", *J. Opt. Soc. Am. B*, **6**, pp. 2194–2205, (1989).
93. G. J. Dick, J. D. Prestage, C. A. Greenhall, and L. Maleki, "Local oscillator induced degradation of medium-term stability in passive atomic frequency standards", *22nd PTTI, NASA CP 3116*, pp. 487–508, (1990).
94. R. Barillet, V. Giordano, J. Viennet, and C. Audoin, "Limitation of the clock frequency stability by the interrogation frequency noise: Experimental results", *IEEE Trans. Instrum. Meas.*, **42**, no. 2, pp. 276–280, Apr. (1993).
95. G. D. Rovera, G. Santarelli, and A. Clairon, "Frequency synbook chain for the atomic fountain primary frequency standard", *IEEE Trans. Ultrason. Ferroelectr. Freq. Control*, **43**, no. 3, pp. 354–358, May (1996).
96. R. Boudot, S. Guérandel, and E. de Clercq, "Simple-design low noise NLTL-based frequency synthesizers for a CPT Cs clock", *IEEE Trans. Instrum. Meas.*, **58**, no. 10, pp. 3659–3665, Oct. (2009).
97. F. R. Martinez, M. Lours, P. Rosenbusch, F. Reinhard, and J. Reichel, "Low phase noise frequency synthesizer for the trapped atom clock on a chip", *IEEE Trans. Ultrason. Ferroelectr. Freq. Control*, **57**, no. 1, pp. 88–93, Jan. (2010).
98. R. Boudot, "Simple design low noise NLTL-based frequency synthesizers for a CPT Cs clock," *IEEE Trans. on Instr. and Meas*, (2009).
99. Picosecond Pulse Labs, "A new breed of comb generators featuring low phase noise and low input power," *Microwave Journal*, **49**, no. 5, p. 278, May (2006).
100. R. C. Mockler, "Atomic beam frequency standards," *Advances Electron. Electron. Phys.*, **15**, (1961).
101. R. J. Carpenter, E. G. Beaty, P. L. Bender, S. Saito, and R. O. Stone, "Prototype Rb vapour frequency standard," *IRE Trans. Instrum.*, 1–9, pp. 132–135, (1960).

102. R. A. Baugh and L. S. Cutler, "Precision frequency sources," *Microwave J.*, **13**, pp. 430–456, (1970).
103. G. H. Heilmeier, "Personal communications: Quo vadis," *Proc. IEEE Int. Solid-State Circ. Conf. Dig.*, pp. 24–26, Feb. (1992).
104. J. L. Wang, T. Tsuchiya, H. Takeuchi, et al., "A miniature ultra-low power 2.5 GHz down converter IC for wireless communications," *NEC Res. & Develop.*, **35**, pp. 46–50, Jan. (1994).
105. C. Denig, M. McCombs, J. Ortiz, et al., "A silicon bipolar monolithic down converter for the commercial motorola global positioning system receiver," *1993 IEEE Bipolar Circuits and Technol. Meet.*, pp. 84–87, Oct. (1993).
106. B. Khabbaz, A. Douglas, J. De Angelis, et al., "A high performance 2.4 GHz transceiver chip-set for high volume commercial applications," *1994 IEEE Microwave and Millimeter-Wave Monilithic Circuit Symp.*, pp. 11–14, May (1994).
107. M. S. Wang, M. Carriere, P. O'Sullivan, and B. Maoz, "A single-chip MMIC transceiver for 2.4 GHz spread spectrum communications," *IEEE Microwave and Millimeter-Wave Monolithic Circuit Symp.*, pp. 19–22, May (1994).
108. M. Funabashi, T. Inoue, K. Ohata, K. Maruhashi, K. Hosoya, M. Kuzuhara, K. Kanakawa, and Y. Kobayashi, "A 60 GHz MMIC stabilized frequency source composed of a 30 GHz DRO and a doubler," *IEEE MTT-S Int. Microw. Symp. Dig.*, pp. 71–74, (1995).
109. C. Yao and J. Xu, "A D-band frequency doubler with GaAs Schottky varistor diodes," *International Journal of Electronics*, **97**, no. 12, pp. 1449–1457, Dec. (1997).
110. J. Zhang and A. V. Raeisaenem, "Improving the CAD of SRD multiplier", *IEEE MTT-S Int. Microw. Symp. Dig.*, **3**, pp. 1767–1770, June (1996).
111. J. Zhang and A. V. Raeisaenem, "A survey on step recovery diode and its applications," Radio Lab., Helsinki Univ. Technol., Helsinki, Finland, Rep. S, **208**, Sept. (1994).
112. A. V. Khramov and V. A. Shchelokov, "The design of microwave transistor multipliers," *Radiotekhnica*, no. 9, pp. 23–25, (1987).
113. A. Z. Venger, A. N. Ermak, and A. M. Yaki,meko, "Simple transistor frequency multiplier," *PRbory i Tekhnika Eksperimenta*, no. 3, pp. 143–144, May–June, (1979).
114. R. H. Johnston and A. R. Boothroyd, "High-frequency transistor frequency multipliers and power amplifiers," *IEEE J. Solid-state Circ.*, SC-7, pp. 81–89, Feb. (1972).

115. D. M. Klymyshyn and Z. Ma, "High order frequency multiplier chain for 28 GHz applications," *IASTED Int. Wireless and Optical Communications Conf.* Banff, Canada, pp. 99–101, June (2011).
116. C. Rauscher, "High-frequency doubler operation of GaAs field-effect transistors," *IEEE Trans. Microw. Theory Tech.*, **31**, no. 6, pp. 462–473, (1983).
117. Y. Iyama, A. Iida, T. Takagi, and S. Urasaki, "Second-harmonic reflector type high-gain FET frequency doubler operating in K-band," *IEEE MTT-S Int. Microw. Symp. Dig.*, pp. 1291–1294, (1989).
118. D. G. Thomas, Jr. and G. R. Branner, "Single-ended HEMT multiplier design using reflector networks," *IEEE Trans. Microw. Theory Tech.*, **49**, no. 5, pp. 990–993, (2001).
119. J. Gavan, and A. Peled, "Low-power passive frequency doublers of high efficiency using varactor diodes", *International Journal of Electronics*, **63**, no. 6, pp. 1011–1019, June (1990).
120. M. S Gupta, R. W. Laton, and T. T. Lee, "Frequency multiplication with high-power microwave field-effect transistors," *IEEE MTT-S Int. Microw. Symp. Dig.*, pp. 498–500, Apr. (1979).
121. J. J. Pan, "Wideband MESFET Frequency Multiplier," *IEEE MTT-S Int. Microw. Symp. Dig.*, pp. 306–308, June (1978).
122. P. T. Chen, C. T. Lee, and P. H. Wang, "Dual-gate GaAs FET as a frequency multiplier at Ku-band," *IEEE MTT-S Int. Microw. Symp. Dig.*, pp. 309–311, June (1978).
123. A. Gopinath and J. B. Rankin, "Single-gate MESFET frequency doublers," *IEEE Trans. Microw. Theory Tech.*, **30**, no. 6, pp. 869–875, 1982, ISSN 0018-9480.
124. S. A. Maas, "Non-linear microwave and RF circuit," Norwood: Artech House, pp. 582, (2003), ISBN 1-58053-484-8.
125. E. O'Ciardha, S. U. Lidholm, and B. Lyons, "Generic-device frequency-multiplier analysis-a unified approach." *IEEE Trans. Microw. Theory Tech.*, **48**, no. 7, pp. 1134–41, ISSN 0018-9480 (2000).
126. H. Fudem and E. C. Niehenke, "Novell millimeter wave active MMIC triplers.," *IEEE MTT-S Int. Microwave Symp.* Baltimore, pp. 387–390, (1998).
127. C. H. Rauscher, "High-frequency doubler operation of GaAs field-effect transistors," *IEEE Trans. Microw. Theory Tech.*, **31**, no. 6, pp. 462–473, ISSN 0018-9480, (1983).
128. E. Camargo, "Design of FET frequency multipliers and harmonic oscillators," Norwood: Artech House, pp. 215, (1998), ISBN 0-89006-481-4.

129. M. Borg and G. R. Branner, "Novel MIC bipolar frequency doublers having high gain, wide bandwidth and good spectral performance," *IEEE Trans. Microw. Theory Tech.*, **39**, no. 12, pp. 1936–1946, ISSN 0018-9480, (1991).
130. D. G. Thomas and G. R. Branner, "Optimization of active microwave frequency multiplier performance utilizing harmonic terminating impedances," *IEEE Trans. Microw. Theory Tech.*, **44**, no. 12, pp. 2617–2624, ISSN 0018-9480, (1996).
131. J. E. Johnson, G. R. Branner, and J. P. Mima, "Design and optimization of large conversion gain active microwave frequency triplers," *IEEE Microw. Wirel. Compon. Lett.*, **15**, no. 7, pp. 457–459, ISSN 1531-1309 (2005).
132. S. A. Maas, *Non-linear Microwave Circuits*, Norwood, MA: Artech House, (1988).
133. P. Antognetti and G. Massobrio, *Semiconductor Device Modelling with SPICE*, (2nd ed.). New York, NY: McGraw-Hill, (1993).
134. R. S. Muller and T. I. Kamins, *Device Electronics for Integrated Circuits*, New York, NY: John Wiley & Sons, Inc., (1982).
135. R. Anholt, Electrical and Thermal Characterization of MESFET'S, HEMT's and BT's, Norwood, MA: Artech House, (1995).
136. J. Zhang and A. Raisanen, "A survey on step recovery diode and its applications," *Report S 208*, Radio Laboratory, Helsinki University of Technology, Finland, ISBN 951-22-2270-1, ISSN 1237-4938, Sept. (1994).
137. A. Duzdar and G. Kompa, "Applications using a low-cost baseband pulsed microwave radar sensor," in *Proc. 18th IEEE Instrumentation and Measurement Technology Conf.*, pp. 239–243, (2001).
138. J. S. Lee and C. Nguyen, "Uniplanar picosecond pulse generator using step-recovery diode," *Electron. Lett.*, **37**, no. 8, pp. 504–506, Apr. (2001).
139. J. Han and C. Nguyen, "A new ultra-wideband, ultra-short monocycle pulse generator with reduced ringing," *IEEE Microw. Wireless Compon. Lett.*, **12**, no. 6, pp. 206–208, Jun. (2002).
140. K. Madam and C. S. Aitchison, "A 20 GHz mcrowave sampler," *IEEE Trans Microwave Theory Tech.*, **140**, pp. 1960–1963, Oct. (1992).
141. W. C. Wlìteley, W. E. Kunz, and W. J. Aankam "50GHz sampler hybrid utilizing a small shockline and an internal SRD," In *1991 IEEE MTT-S Int Microwa Symp. dig.*, Boston MA, pp. 895–898, (1991).
142. S. R. Gibson, "Gallium arsenide lowers cost and improves performance of mcrowave counters," *Hewlett-Packard J.*, pp. 4–10, Feb. (1986).

143. A. N. Pergande, "One watt. W-band transmtter," *in 1994 IEEE MTT-S Int. Mzcrowave Symp Dig*, San Diego, CA, pp. 305–308, June (1994).
144. Anfu Zhu, Fu Sheng, Anxue Zhang, "An Implementation of Step Recovery Diode-Based UWB Pulse Generator", *Proceedings of IEEE International Conference on Ultra-Wideband (ICUWB2010)*, (2010).
145. J. R. Andrews, "Picosecond Pulse Generation Techniques and Pulser Capabilities," icosecond Pulse Labs., Boulder, CO, Applicat. Note AN-30419, Nov. (2008).
146. Howard Charles Reader, Dylan, F. Williams, Paul, D. Hale, and Tracy, S. Clement, Comb-generator characterization," *IEEE Trans. Microwave Theory & Tech.*, **56**, no. 2, pp. 515, Feb. (2008).
147. D. L. Hedderty "An analysis of a circuit for the generation of high-order harmonics using an ideal non-linear capacitor. *IRE Tran. Electron Devices*, ED-9, pp. 484–491, November (1962).
148. S. M. Krakauer, "Harmonic generation, rectification and lifetime evaluation with the step-recovery diode", *Proc. Inst. Radio Engrs*, 50, p. 1665, (1962).
149. D. J. Roulston, "Frequency multiplication using a charge storage effect: An analysis for high efficiency, high power operation", *International Journal of Electronics*, **18**, pp. 73–86. April (2007).
150. R. Thompson, "Step-recovery diode frequency multiplier", *Electronics Letters*, **2**, no. 3, p. 117, March (1966).
151. S. Hamilton and R. Hall, "Shunt-mode harmonic genera & usmg step recovery diodes," *Microwave J.*, **10**, pp. 69–78, April (1967).
152. P. Chilvers and K. Foster, "Simulation of a high order charge-storagc-diode multiplier on an analogue Computer". *Electronics Letters* 3, pp. 277–278, June (1967).
153. R. Thompson, "Mismoding in harmonic multipliers," *Electronies Letter*, **3**, pp. 402–403. September (1967).
154. J. Zhang and A. V. Raisanen, " A new model of step recovery diodes for CAD", *Digest of the 1995 IEEE International Symposium on Microwave Theory and Techniques*, Orlando, USA, pp. 1459–146:2. May (1995).
155. G. D. Cormack and A. P. Sabharwal, "Picosecond pulse generator using delay lines," *IEEE Trans.Instrum. Meas.*, **42**, no. 5, pp. 947–948, Oct. (1993).
156. S. Goldman, "Computer aids design of impulse multipliers," *Microwaves & RF*, pp. 101–128, Oct. (1983).
157. S. Akhtarzad, T. R. Rowbotham, and P. B. Johns, "The design of coupled microstrip lines," *IEEE Trans Microwave Theory Tech.*, 23, pp. 486–492, June (1975).

158. P. W. C. Chilvers and K. Foster, "Simulation of a high-order, charge-storage-Diode multiplier on an analogue computer", *Electron. Lett.*, 3, pp. 277–278, June (1967).
159. J. L. Moll, et al., "Physical modelling of the step recovery diode for pulse and harmonic generation circuits," *Proc. IEEE*, **57**, pp. 1250–1259, July (1969).
160. J. Zhang and A. V. Raisanen "Computer-aided design of step recovery diode frequency multipliers," *IEEE Trans. Microwave Theory & Tech.*, **44**, 12, pp. 2612–2616, Dec. (1996).
161. J. Zhang and A. V. Raisanen, "A survey on step recovery diode and its applications," *Report S 208*, Radio Lab., Helsinki Univ of Technol, Sept. (1994).
162. Harmonic generation using step recovery diodes, *Hewlett-Packard Application Note #920*.
163. S. Hamilton and R. Hall, "Physical modelling of the step recovery diode for pulse and harmonic generation circuits," *Proc. IEEE*, **57**, pp. 1250–1259, July (1969).
164. S. A. Maas, Non-linear microwave circuits norwood, MA. Artech House, (1988).
165. J. Han, M. Miao, and C. Nguyen, "Recent development of SRD- and FET-based sub-nanosecond pulse generators for ultra-wideband communications," *IEEE Topical Conference on Wireless Communication Technology*, pp. 441–442, (2003).
166. T. Edwards, Foundations for Microstrip Circuit Design, 2nd ed. Chichester, England, Wiley, (1992).
167. B. A. Syrett, "A Broadband element for microstrip bias or tuning circuits," *IEEE Trans. MTT*, Mtt-28, no. 8, pp. 488–491, August (1980).
168. Y. Y. Wei, P. Gale, and E. Korolkiewicz, "Effects of grounding and bias circuit on the performance of high frequency linear amplifiers," *Microwave Journal*, **46**, no. 2, pp. 98–106, February (2003).
169. A. Bhargava, "Amplifier design made simple". Application Engineer EEs of EDA, Agilent Technologies, Inc. (2006).
170. P. Dixon, "Dampening cavity resonance using absoRber material," *RF Design Magazine*, pp. 16–19, May (2004).
171. W. D. Lee, J. H. Shirley, F. L. Walls, and R. E. Drullinger, "Systematic errors in cesium beam frequency standards introducesd by digital control of the microwave excitation", *Proc. Int. Freq. Cont. Symp.*, pp. 113–117, (1995).
172. C. Audoin, M. Jardino, L. S. Cutler, and R. F. Lacy, "Frequency offset due to spectral impurities in cesium-beam frequency standard," *IEEE Trans. I & M.* IM-27, pp. 325–329, (1978).

173. M. V. Romalis, E. Miron, and G. D. Cates, Pressure broadening of RbD1 and D2lines by 3He, 4He, N2, and Xe:Line cores and near wings, *Physical Review A*, **56**, 4569, (1997).
174. J. H. Shirley, "Some causes of resonant frequency shifts in atomic beam machines-II, The effect of slow frequency modulation on the Ramsey line shape," *J. Appl. Phys.*, **34**, no. 4, part 1, pp. 789–791, Apr. (1963).
175. N. F. Ramsey. "Resonance transitions induced by perturbations at two or more frequencies." *Phy. Rev.*, **100**, no. 4, pp. 1191–1194, Nov.
176. C. Audoin, V. Candelier, and N. Dimarcq. "A limit to the frequency stability of passive frequency standards due to an intermodulation effect," *IEEE Trans. Instrum. Meas.*, **40**, pp. 121–125, Apr. (1991).
177. C. Affolderback, E. Breschi, C. Schori, and G. Mileti 27–30 June 2006, Gas cell atomic clocks for space: New results and alternative schemes, *Proc. International conf. on space optics*, (10567).
178. M. Gharavipour, et al., High performance vapour-cell frequency standards, *J. Phys.: Conf. Ser.*, **723**, 012006C (2016).
179. H. E. Williams, T. M. Kwon, and T. Mellelland, *Proc. 31st Annual Symposium on Frequency Control*, p. 12, (1983).
180. A. F. Podell and L. F. Mueller, U.S. Patent 4,349,798 (1982).
181. H. S. Schweda, G. Busca, and P. Rochat, U.S. Patent 5,387,881 (1995).
182. Y. Shiyu, C. Jingzhon, T. Jianhui, and L. Yaoting, "A kind of magnetron cavity used in Rb atomic frequency standards", *Journal of Semiconductors*, **32**, no. 12 December (2011).
183. T. M. Kwon and H. E. Williams, U.S. Patent 4,495,478 (1985).
184. G. Busca and L. Johnson, U.S. Patent 4,947,137 (1990).
185. W. N. Hardy and L. A. Whitehead, Review of scientific instruments **52**, 213–216, (1981); W. Froncisz and J. S. I-I.e. U.S. Patent 4,446,429 (1984); R. Bien, et al., U.S. Patent 4,633,180 (1986).
186. G. H. Mei and J. T. Liu, *Proc. 1999 Joint Meeting of EFTF and IEEE IFCS*, Besancon, France; p. 601, Apr. (1999).
187. E. Eltsufin., Stern, Avinoam, and Fel, S. *Proc. 45th Annual Symposium on Frequency Control*, pp. 567–571, (1991).
188. T. V. C. T. Chan and H. C. Reader, "Understanding microwave heating cavities", Artech House, Boston, London, (2000).
189. H. Kubo, I. Awai, Y. Ishii, K. Iwashita, and A. Sanada, "Improvement of unloaded Q of image dielectric resonator due to shift of electromagnetic field distribution". 33rd *European Microwave Conference* — Munich (2003).

190. M. N. Pospieszalski, "Cylindrical dielectric resonators and their applications in TEM line microwave circuits," *IEEE Trans, Microwave Theory Tech, MTT-27*, pp. 233–238, Mar. (1979).
191. S. F. Ziuszko and A. Jelenski, "The influence of conducting walls on resonant frequencies of the dielectric resonator," *IEEE Trans. Microwave Theory Tech, MTT-19*, p. 778, Sept. (1971).
192. T. Itoh and R. Rudokas, "New method for computing the resonant frequency of dielectric resonator," *IEEE Trans. Microwave Theory Tech. MTT-25*, pp. 52–54, Jan. (1977).
193. P. Guillon and Y. Garault, "Accurate resonant frequencies of dielectric resonators," *EEE Trans. Microwave Theory Tech., MTT-25*, pp. 916–922, Nov. (1977).
194. S. Moraad, S. Verdeymo, P. Guillon, P. Ulian, and B. Theron, "A new planar type dielectric resonator for microwave filtering", *IEEE MTT-S International*, 3, pp. 307–1314, June (1998).
195. A. F. Podell and L. F. Mueller, U.S. Patent 4,349,798 (1982).
196. H. S. Schweda, G. Busca, and P. Rochat, U.S. Patent 5,387,881 (1995).
197. A. Peter Rizzi, Microwave engineering passive circuits, Prentice-Hall, inc. (1988).
198. D. Kajfez and P. Guillon, Dielectric Resonator, Artech House Dedham, MA, Ch. 4 (1986).
199. D. M. Pozar, Microwave Engineering, John Wiley & Sons, inc. (1988).
200. G. H. Mei and J. T. Liu, *Proc. 1999 Joint Meeting of EFTF and IEEE IFCS*, Besancon, France; p. 601 Apr. (1999).
201. J. Vaneir and C. Audoin, chapter 2, The Quantum Physics of Atomic Frequency Standard Adam Hilger, (1989).
202. C. AffoldeRbach, F. Gruet, D. Miletic, and G. Milleti, in Proceedings of the 7th Symposium on Frequency Standards and Metrology, USA, edited by L. Maleki (World Scientific, 2009), pp. 363–367, 5–11 October (2008).
203. S. Johnson, M. Thomas, and C. Kros, "Membrane capacitance measurement using patch clamp with integrated self-balancing lock-in amplifier," *Pflugers Arch—Eur. J. Physiol.*, **443**, no. 4, pp. 653–663, Feb. (2004).
204. M. Sonnaillon, R. Urteaga, F. Bonetto, and M. Ordonez, "Implementation of a high-frequency digital lock-in amplifier," in *Proc. Can. Conf. Electrical and Computer Engineering*, pp. 1229–1232, May (2005).
205. A. De Marcellis, G. Ferri, M. Patrizi, V. Stornelli, A. D'Amico, C. Di Natale, E. Martinelli, A. Alimelli, and R. Paolesse, "An integrated analog lock-in amplifier for low-voltage low-frequency sensor

inteRFace," in *Proc. 2nd Int. Workshop Advances in Sensors and InteRFace*, pp. 1–5, June (2007).
206. A. Gnudi, L. Colalongo, and G. Baccarani, "Integrated lock-in amplifier for sensor pplications," in *Proc. Eur. Solid-State Circuits Conf.* pp. 58–61, Sep. (1999).
207. C. Azzolini, A. Magnanini, M. Tonelli, G. ChioRboli, and C. Morandi, "Integrated lock-in amplifier for contactless inteRFace to magnetically stimulated mechanical esonators," in *Proc. 3rd Int. Design and Technology of Integrated Systems Nanoscale Era*, pp. 1–6, Mar. (2008).
208. G. Ferri, P. De Laurentiis, A. D'Amico, and C. Di Natale, "A low voltage integrated CMOS analog lock-in amplifier prototype for LAPS applications," *Sensors Actuators A: Phys.*, **92**, no. 1, pp. 263–272, Aug. (2001).
209. Stanford Res. Sys., Model SR830 DSP lock-in amplifier. (2006).
210. H. Goovaerts, J. Faes, E. Raaijmakers, and R. Heethaar, "A wideband high common-mode rejection ratio amplifier and phase-locked loop demodulator for multi-frequency impedance measurement," *Med. Biol. Eng. Comput.*, **36**, no. 6, pp. 761–767, Nov. (1998).
211. M. L. Meade, Lock-in Amplifiers: Principles and Applications, Peter Peregrinus Ltd., (1983).
212. Lock-in amplifiers, appl. notes, Stanford Res. Sys., data sheets, (1999).
213. W. Demtroder, Laser spectroscopy: Basic concepts and instrumentation. Springer Verlag, Berlin. (1981).
214. A. Corney, Atomic and laser spectroscopy. Clarendon Press, Oxford, New York. (1977).
215. A. De Marcellis, G. Ferri, M. Patrizi, V. Stornelli, A. D'Amico, C. Di Natale, E. Martinelli, A. Alimelli, and R. Paolesse, "An integrated analog lock-in amplifier for ow-voltage low-frequency sensor inteRFace," in *Proc. 2nd Int. Workshop Advances in Sensors and InteR-Face*, pp. 1–5, Jun. (2007).
216. A. Gnudi, L. Colalongo, and G. Baccarani, "Integrated lock-in amplifier for sensor applications," in *Proc. Eur. Solid-State Circuits Conf.*, pp. 58–61, Sep. (1999).
217. C. Azzolini, A. Magnanini, M. Tonelli, G. ChioRboli, and C. Morandi, "Integrated lock-in amplifier for contactless inteRFace to magnetically.
218. A. Gnudi, L. Colalongo, and G. Baccarani, Integrated lock-in amplifier for sensor applications, *SolidState Circuits Conference*, pp. 58–61, Sept. (1999).

219. A. D'Amico, et al., "Low-voltage low-power integrated analog lock-in amplifier for gas sensor applications", Sens. Actuators B: Chem., In Press, 2009. **767**, Nov. (1998).
220. S. Franco, "Design with operation ampifiers and analog integrated circuits", 3rd ed., McGraw-Hill, New York, (2002).
221. G. Ng, B. Lai, P. Liu, S. P. Voinigescu, "1GHz opamp-based bandpass filter," *Silicon Monolithic Integrated Circuits in RF Systems*, Digest of Papers. 2006, Topical Meeting on 18–20, pp. 4, Jan. (2006).
222. R. H. Hamstra and P. Wendland," Noise and frequency response of silicon photo-diode operational amplifier combination" *Appl. Opt.*, **11** 1539–47, (1972).
223. J. A. Ringo and P. O. Lauritzen, "1/f noise in uniform avalanche diodes", *Solid State Electron.*, **16**, 327–8, (1973).
224. F. N. Hooge, "1/f noise sources", *IEEE Trans. Electron Devices*, **41**, 1926–35, (1993).
225. X. Zhao, M. J. Deen, and L E Tarof, "Low frequency noise in separate absorption,grading, charge and multiplication (SAGCM) avalanche photo-diodes" *Electron. Lett.*, **32**, 250–2, (1996).
226. T. Lynch, W. Riley, and J. Vaccaro, "The Testing of Rb Frequency Standards," *Proc. 43rd Ann. Symp. on Freq. Control*, pp. 257–262, May (1989).
227. H. Hellwig, "Environmental Sensitivities of Precision Frequency Sources," *Proc. 3rd European Time and F'requency Forum*, pp. 5–10, March (1989).
228. S. Pancharatnam, "Light shifts in semi classical dispersion theory", *J. Opt. Soc. Am.*, **56**, pp. 1636, (1966).
229. B. S. Mathur, H. Tang, and W. Happer, "Light shifts in the alkali atoms", *Phy. Rev.*, **171**, pp. 11–19, (1968).
230. S. Knappe, V. Shah, P. D. D. Schwindt, L. Hollberg, J. Kitching, L.-A. Liew, and J. Moreland, *Appl. Phys. Lett.*, **85**, 1460, (2004).
231. W. J. Riley, *IEEE Trans. Ultrason. Ferroelectr. Freq. Control* **39**, 232, (1992).
232. Y. Y. Jau, A. B. Post, N. N. Kuzma, A. M. Braun, M. V. Romalis, and W. Happer, Intense, narrow atomic-clock resonance", *Phys. Rev. Lett.*, **92**, 10801, (2004).
233. B. C. Grover, E. Kanegsberg, J. G. Mark, and R. L. Meyer, U.S. Patent no. 4, **157**, 495A, 5 June (1979).
234. K. F. Woodman, P. W. Franks, and M. D. Richards, "The nuclear magnetic resonance gyroscope: A review", *J. Navigation*, **40**, pp. 366–384, September (1987).

235. P. Härle, G. Wäckerle, and M. Mehring, *Appl. Magnetic. Resonance.* **5**, 207–220, (1993).
236. A. K. Thomas," Magnetic shielded enclosure design in the DC and VLF region", *IEEE Trans. Electromagn. Compat.*, **10**, pp. 142–152, March (1968).
237. D. Schwindt, S. Knappe, V. Shah, L. Hollberg, J. Kitching, L.-A. Liew, and J. Moreland, "A chip-scale atomic magnetometer", *Appl. Phys. Lett.*, **85**, 6409–6411, (2004).
238. P. D. Schwindt, B. Lindseth, S. Knappe, V. Shah, and J. Kitching, "A chip-scale atomic magnetometer with improved sensitivity using the Mx technique", *Appl. Phys. Lett.*, **90**, 081102, (2007).
239. A. Mager, *IEEE Trans. Magn.* **6**, pp. 67–75, May, (1970).
240. D.U. Gubser et al., *Rev. Sci. Instrum.* **50**, 751, (1975).
241. Bikash Ghosal, G. M. Saxena design verification model of Rb frequency standard for space, *Journal of Modenn Physics*, **5**, pp. 128–135, (2014).
242. A. Kastler, "Optical methods of atomic orientation and of magnetic resonance," *J. Opt. Soc. Amer.*, **47**, pp. 460–465, June, (1957).
243. E. Arimondo, Progress in Optics XXXV, 259–354, (1996).
244. W. Happer, "Optical pumping", *Rev. Mod. Phys.*, **44**, pp. 169–249, (1972).
245. A. Bloom, "Optical pumping," *Sci. Am.*, **203**, 72–80, (1960).
246. T. Colbert and J. Huennekens, "Radiation trapping under conditions of low to moderate line center optical depth, *Phys. Rev. A*, 41, pp. 6145–6153, (1990).
247. J. Vanier and C. Audoin, Chapter 7, pp. 1312–7, The Quantum Physics of Atomic Frequency Standards. New York: Hilger, (1989).
248. G. Missout and J. Vanier, Pressure and temperature coefficients of more commonly used buffer gases in Rb frequency standards, *IEEE Trans on I & M*, **24**(2) pp. 180–184, (1975).
249. J. Vanier, R. Kunski, N. Cyr, J. Y. Savard, and M. Têtu, "On hyperfine frequency shifts caused by buffer gases: Application to the optically pumped passive Rb frequency standard," *J. Appl. Phys.*, **53**, p. 5387, (1982).
250. J. Vanier and L. G. Bernier, "On the signal-to-noise ratio and short-term stability of passive Rb frequency standards", *IEEE Transactions on Instrumentation and Measurement*, IM-30, No. 4, pp. 277–282, Dec. (1981).
251. J. Vanier, C. Audoin, Chapter 3, "The Quantum Physics of Atomic Frequency Standard", Bristol, Philadelphia: Adam Hilger, pp. 1257–1364, (1989).

252. C. M. Klimcak, M. Huang, and J. C. Camparo, Alkali metal consumption by discharge lamps fabricated from GE-180 aluminosilicate glass, Joint Conference of the EEE International Frequency Control Symposium & the European Frequency and ime Forum, (2015).
253. Antone. Ivanov, Christoph Affolderbach, Gaetano mileti and Anja k. Skrivervik, Design of atomic clock cavity based on a loop-gap geometry and modified boundary conditions, *International Journal of Microwave and Wireless Technologies*, **9**(7), 1373–1386, (2017).
254. J. E. Thomas, P. R. Hemmer, S. Ezekiel, C. C. Leiby, H. Picard, and R. Willis, *Phys. Rev. Lett.*, **48**, p. 867 (1982).

Index

^{85}Rb filter cell, 11, 18
^{87}Rb absorption cell, 11, 18, 19
^{87}Rb bulb, 11, 19

Allan deviation, 109, 209
Analog frequency synthesis, 258
Atomic resonance line, 16
Atomic transitions, 1

Band-pass filter, 200
Base plate, 15, 19
Base plate, 251
BeiDou Satellite Navigation System, 3
Bibliography, 269
Bifilar, 14
Boltzmann distribution, 46
Breit–Rabi splitting, 44
Buffer gas, 14

Calorimeter, 251
capacitance gauge, 244
Cavity pulling, 19, 21, 181
Cesium atomic clock, 2, 28
coherent population trapping, 268
Colpitt/Clapp oscillator, 15
Compass, 3
CPT, 17
crystal filter, 265

D1 and D2 lines, 47
Delyiannis–Friend filter, 201
Design Verification Model DVM, 37
diaphragm valve, 244
Dicke Effect, 17

dielectric loss, 170
Differential Calorimeter, 252
Diode laser, 13
Direct Digital Synthesizer, 267
dither signal, 197
Doppler broadening, 14
double-resonance, 47

Eigenvalues, 44
Electronic Frequency Control, 31
Electronic Package, 14
Engineering Thermal Model (ETM), 38
Error signal, 32

C-field, 14
field effect transistors, 112
fluorescent photon, 248
Frequency stability, 1
Frequency synthesizer, 28

Galileo, 3, 13, 48
Generation–Combination Noise, 25
GLONASS, 3, 48
GPS, 1
Gummel–poon model, 120
gyro-magnetic ratio, 44

Hamiltonian, 44
harmonic balance simulation, 59
He permeability, 27
Helium permeability, 253
HFFS, 30
high electron mobility transistors, 112

Index

High-pass filter, 200
Hyperfine levels, 1, 19

IFT, 13, 18–20, 27
Integrator, 255
International System of Units, 8
inter-stage-conjugate-matching, 153
IRNSS, 4
ISRO, 4

Johnson Noise, 25

Larrmor precession, 44
Light shift, 15, 19–21
Lock-in-amplifier, 31, 37, 188
Loop-gap microwave cavity, 268
Low-pass filter, 204

magnetic dipole transitions, 44
Magnetic field solenoid, 250
Magnetic shield, 214
H-Maser, 2, 193
micro-strip lines, 153
Microwave cavity, 23, 24, 29, 30, 193
Microwave-optical double resonance, 47

National Bureau of Standards, 8
NAVSTAR GPS, 2
NIST, 8
Notch Filter, 256
NPLI, 48
nuclear magneton, 44
numerically controlled oscillator, 259

OCXO, 1, 28, 33, 193, 210
Optical pumping, 17, 19, 46, 47

S-parameter, 55
permeability, 168
permittivity, 168
phase modulation, 157
Phase shifter, 202
Photon Noise, 24
photo-voltaic cell, 12

Physics package, 13, 37
PID temperature controller, 105
Plank's constant, 249
Population inversion, 11
population inversion, 46
Precision OCXO, 109
Pressure shift, 19

quantization axis, 43
Quartz crystals, 7
Quasi-Zenith Satellite System, 4
Quenching, 247

radiation shielding, 22
Radiation trapping, 17
Rb atomic clock, 1, 19, 22, 24, 27, 40, 44, 47
Rb bulb and absorption cell filling, 244
Rb lamp, 27, 51
Regional navigation satellite system, 2, 4
RF power shift, 19, 21
RF Synthesizer, 106
Ring-mode, 16

Satellite navigation system, 1, 4
Second harmonic, 14
Self-reversal, 16
SFT, 13, 18–20, 27
Short term stability, 1
short term stability, 233
signal-to noise ratio, 190
Smith Chart, 177
Space and Radiation effect, 19, 22
Space Application Centre, 4
spin-orbit coupling, 42
SRD, 29
Stark shift, 20
Stefan–Boltzmann constant, 99
Step recovery diodes, 130
Sundials, 7
SWG, 250
synchronous orbits, 2

Temperature coefficient, 17
Temperature controller, 26
Temperature sensitivity, 231
The space qualified Rb clocks, 48
thermal conductivity, 98
Thermal Mathematical Model, 88
Thermistor, 17
Thermo-vacuum test, 98, 100
trans-impedance amplifier, 197
Trilateration, 7

UHV, 27

VCSEL, 13
VCXO, 1, 11, 31, 193, 210
Vector Network Analyser, 52

Zeeman splitting, 43

μ-metal material, 250
π-transitions, 44
σ-transitions, 44